VENTILATION FOR
ENVIRONMENTAL
TOBACCO SMOKE

VENTILATION FOR
ENVIRONMENTAL
TOBACCO SMOKE

By

BRIAN A. ROCK, PH.D., P.E.

AMSTERDAM • BOSTON • HEIDELBERG • LONDON
NEW YORK • OXFORD • PARIS • SAN DIEGO
SAN FRANCISCO • SINGAPORE • SYDNEY • TOKYO
BUTTERWORTH–HEINEMANN IS AN IMPRINT OF ELSEVIER

ELSEVIER

But erworth-Heinemann is an imprint of Elsevier
30 Corporate Drive, Suite 400, Burlington, MA 01803, USA
Linacre House, Jordan Hill, Oxford OX2 8DP, UK

ASHRAE
1791 Tullie Circle, N.E.
Atlanta, Georgia 30329-2305
www.ashrae.org

Permissions may be sought directly from Elsevier's Science & Technology Rights Department in Oxford, UK: phone (+44) 1865 843830, fax: (+44) 1865 853333, E-mail: permissions@elsevier.com. You may also complete your request on-line via the Elsevier homepage (http://elsevier.com), by selecting "Support & Contact" then "Copyright and Permission" and then "Obtaining Permissions."

Recognizing the importance of preserving what has been written, Elsevier prints its books on acid-free paper whenever possible.

Library of Congress Cataloging-in-Publication Data
Application submitted.

British Library Cataloguing-in-Publication Data
A catalogue record for this book is available from the British Library.

ISBN 13: 978-0-12-370886-1

ISBN 10: 0-12-370886-9

For information on all Butterworth-Heinemann
publications visit our Web site at www.books.elsevier.com

05 06 07 08 09 10   10 9 8 7 6 5 4 3 2 1

Printed in the United States of America

CONTENTS

1 Introduction 1

2 What Is ETS? 7

3 Indoor Environmental Quality 21

4 Ventilation Rates for ETS 43

5 ETS Design Issues 67

6 Applications 117

7 Summary 171

   Nomenclature 175

   Bibliography 177

   About the Author 185

   Index 187

   Comment Sheet 207

1

INTRODUCTION

This book was written to help HVAC system designers and others address a very specific ventilation concern—odor and irritation control when environmental tobacco smoke (ETS) is present. In recent years, much has been learned through research and experience about improving ventilation of buildings, and specific measures for doing so have evolved. But the issue of how to handle ETS in buildings persists, and HVAC designers have had little guidance available to them. This book is intended to fill this need. Its purpose is not to present specific requirements, as would a code, standard, or guideline, but instead to provide relevant design information.

I hope that this book will serve as a practical introduction for those who are just beginning to address ETS design issues. The book can also serve as a refresher for those who are more experienced with the topic and to stimulate further discussion and research. Reviewing the book may also be beneficial for architects, building owners, system operators, occupants, and code officials so that they can better communicate with HVAC designers on this and related topics. Others may find this book useful as an introduction to some of the technical issues and design challenges. Some suggested additional readings are listed at the end of this chapter.

1.1. ORGANIZATION OF THIS DESIGN GUIDE

This book is organized with the general topics first and then more specific ones later. This first chapter presents background information on HVAC design and system operation. The second chapter introduces the primary topic of concern: environmental tobacco smoke. In the third chapter, many factors that affect indoor environmental quality are reviewed. Various methods for estimating needed ventilation rates are found in Chapter 4. Some specific ways for addressing tobacco smoke with engineering and architectural measures are then presented in Chapter 5. Design issues and examples for various applications are found in Chapter 6, and a summary follows in Chapter 7. A reference list of cited materials, a nomenclature listing, an index, and a comment sheet then complete this book.

1.1.1. Units Used

The primary units used in this book are from the inch-pound (I-P) system, but the International System of Units (SI) equivalents are given in parentheses. As of this writing, use of I-P units is still dominant for HVAC projects constructed within the United States and for the equipment manufactured here. Familiarity with SI, and a transition to its use, is occurring very slowly in the United States, however. Some conversion factors most applicable to the materials presented in this book are:

$$1 \text{ ft} = 0.3048 \text{ m}$$
$$1 \text{ ft/min (FPM)} = 0.005079 \text{ m/s}$$
$$1 \text{ ft}^3/\text{min (CFM)} = 0.4719 \text{ L/s (LPS)} \approx 0.5 \text{ LPS}$$

$$1 \text{ in.w.g.} = 248.8 \text{ Pa}$$
$$3.412 \text{ Btu/h} = 1 \text{ W}$$
$$746 \text{ W} = 1 \text{ hp}$$
$$(5/9) \cdot (^\circ\text{F} - 32) = {}^\circ\text{C}$$

1.2. HVAC DESIGN

Heating, *ventilating*, and *air-conditioning* (*HVAC*) system design involves both art and science. To be successful, an HVAC designer should become well versed in the phenomena affecting system performance. But HVAC designers are a diverse group with a variety of educational backgrounds and experience, and HVAC systems range from the very simple to highly complex. There are often many possible successful solutions for the same HVAC project.

Most codes in the United States require that large engineering projects, including new or significantly revised HVAC systems, be designed under the direct supervision of a registered professional engineer (P.E.). In some firms, this licensed engineer will do the actual HVAC design or be part of or lead a team that includes others such as engineers-in-training, technicians, drafters, and support staff. Many HVAC designers are not P.E.s but should have considerable training and design experience.

Building codes or regulations often do not require that a seal from a registered professional engineer be obtained for designs of smaller buildings' HVAC systems; for example, single-family home systems are usually exempt. The HVAC systems for these small residential buildings are often selected by contractors, builders, or owners, for example, and frequently don't have provisions for admitting outside air through their HVAC systems. Small or older commercial buildings often have residential-like heating and/or air-conditioning systems and may lack provisions for forced ventilation as well, especially where building codes are, or were, weak or poorly enforced.

HVAC systems for larger buildings are normally selected by engineers or designers, and need to comply with the ventilation requirements in building codes and standards. They typically do so through the admission of outside air through the systems, the exhaust of used air, and possibly with enhanced air cleaning and energy recovery provisions. Such complex systems are commonly employed for commercial, institutional, and high-rise residential buildings.

1.3. OPERATION AND MAINTENANCE

As with other types of machinery, large HVAC equipment and systems require operation by trained individuals, who must be supplied with the proper tools, materials, and other resources. The systems need preventive maintenance, such as regular bearing lubrication and filter replacement, and also event-driven repairs, such as replacement of burned-out motors or fixing fluid leakages. Eventually the systems will need partial or total replacement as they fail, become insufficient in capacity, or are found to be technically, economically, or otherwise functionally obsolete.

Even when HVAC systems are properly selected, installed, and commissioned, poor thermal comfort, unacceptable indoor air quality, or increased noise, for example, can result from improper operation or maintenance. Building owners and operators need to make sure that their personnel have adequate training, resources, and oversight so that potential comfort, air quality, and other HVAC-related problems can be avoided, or, if they occur, can be corrected rapidly. Owners should prepare operation and maintenance (O&M) manuals for their HVAC systems so that methods and hands-on experiences can be passed on as personnel change. But when HVAC-related problems, including ETS concerns, are not readily addressed by in-house workers, owners should retain the services of HVAC engineering consultants and outside repair firms to quickly and efficiently identify and resolve problems.

1.4. RELATED READINGS

This book for HVAC engineers, designers, and others is intended only as an introduction to ETS-related odor and irritant control. Many physical, thermal comfort, and indoor air-quality needs should influence your engineering designs. If more in-depth information is required than provided here, you are encouraged to examine the original documents listed in the References section at the end of this book, and in the publications that follow.

1.4.1. Must-Have References

The four-volume *ASHRAE Handbook* includes many chapters on highly relevant topics, such as ventilation and infiltration (Chapter 26), space air diffusion (Chapter 32), duct design (Chapter 34), and indoor-air issues (e.g., Chapters 9, 12, and 13) (*Fundamentals* volume, ASHRAE 2001). An excellent chapter on fans also appears in the *Handbook* (Chapter 18, *HVAC Systems and Equipment* volume, ASHRAE 2004). Even more information on fans, which are needed to ensure proper and consistent airflow rates, can be found in various publications from the Air Movement and Control Association (e.g., AMCA 1990), manufacturers, and elsewhere.

ANSI/ASHRAE Standard 62.1-2004, Ventilation for Acceptable Indoor Air Quality (ANSI/ASHRAE 2004a), addresses commercial, institutional, and high-rise residential buildings and is the standard for determining their general ventilation rates. While earlier versions of this standard contained minimum outdoor air requirements for smoking-permitted spaces, it currently does not. Standard 62.1 is under continuous review, so all adopted *addenda*, and the *official interpretations*, should be obtained along with the main document; it was renumbered recently from 62 and the previous version was 62-2001 plus its addenda. A new, separate Standard 62.2 is concerned with ventilation in smaller residential buildings and has recently been published (ANSI/ASHRAE 2004b). This ETS book does not address these low-rise residences, such as single-family houses.

SMACNA's (1990) and ASHRAE's forthcoming duct design manuals are invaluable tools for sizing, selection, and specification of ductwork. These ducts carry the needed conditioning and ventilating air and can be used to remove contaminants from buildings.

The ASHRAE *Designer's Guide to Ceiling Based Air Diffusion* (Rock and Zhu 2002) covers the thermal, acoustical, and air quality aspects of selecting traditional overhead supply air outlets and return air inlets, and because its lead author also wrote this book, much of the introductory material is similar. However, the specifics needed for placing diffusers, and then how to evaluate their performance, are presented, via discussions and detailed examples, in the 2002 publication by Rock and Zhu. ASHRAE's new design guide for underfloor air distribution (Bauman 2003) provides similar information but for in-floor rather than ceiling-based air supply. Chen and Glicksman's (2003) book addresses displacement air diffusion from a more general point of view.

1.4.2. Related Publications

ANSI/ASHRAE Standard 55-2004, Thermal Environmental Conditions for Human Occupancy (ANSI/ASHRAE 2004c) discusses a significant factor in indoor environmental quality—thermal comfort. HVAC measures that address ETS alone could unintentionally reduce thermal comfort.

The *Cold Air Distribution System Design Guide* (Kirkpatrick and Elleson 1996), published by ASHRAE, considers low-flow rate, high temperature–difference air supply. This type of system, usually installed overhead, may or may not be appropriate for ETS work due to its reduced air-flow rates as compared with more conventional, moderate temperature–difference air distribution systems.

The *Principles of Smoke Management Systems* (Klote and Milke 2002), also published by ASHRAE, addresses fire-event smoke production and removal. There are many similarities between designing for fire-related smoke and ETS, and fire-protection issues should always be considered in the design of any HVAC system.

1.5. OTHER READINGS

A largely nonnumerical introduction to HVAC design is the *Air-Conditioning Systems Design Manual* (ASHRAE 1993). Heating and cooling load calculations for ventilation and thermal comfort are covered in several *Fundamentals* volume chapters of the *ASHRAE Handbook* (2001) and in the *ASHRAE Load Calculation Manual (LCM)*. The current and now third version of the *LCM* was authored by Pedersen et al. (1998), and a new version is under development.

Many publications are available about the health risks of exposure to tobacco smoke and other aspects of tobacco use. Some examples are publications from NCI, the National Toxicology Project, and the CDC Office on Smoking and Health (e.g., EPA 1992; NIH 1999; DHHS 2002; IARC 2002).

As periodically revised and expanded editions of this book are possible, your suggestions for such are encouraged. A comment sheet is provided at the end of this book for this purpose.

2

WHAT IS ETS?

This short chapter introduces tobacco, ways that tobacco is used, many health consequences, what comprises *environmental tobacco smoke* (*ETS*), how people are exposed, and how we characterize the products of combustion for design purposes. Odors and irritants are also addressed. With these topics introduced, later chapters will discuss more specifically the engineering, architectural, and other measures needed for ETS control.

2.1. TOBACCO

The various tobacco plants are native to the Americas and most likely originated in the high mountains of the Andes. The word *tobacco* seems to be of Caribbean origin, and many consider the plant to be a weed. European explorers observed the early North, Central, and South American inhabitants using tobacco; the western settlers who followed then cultivated and exported tobacco and introduced its methods of use to Europe, Turkey, and beyond. Asians are now some of the greatest tobacco consumers. Usage rates of men tend to greatly exceed that of women in most parts of the developing world (e.g., WHO 2002; Gately 2001).

Ancient Americans and early western tobacco sellers and users claimed positive medicinal effects from tobacco. These claims were largely around the included stimulant *nicotine* and the antiseptic and analgesic effects of the plant. However, nicotine in larger quantities was also observed to be a poison, and tobacco was used as an early insecticide (Gately 2001).

Tobacco was, and still is, grown in the Americas and elsewhere as a cash crop. George Washington was a tobacco farmer before changing to other crops later in life as his fields became depleted of the needed nutrients. In early North American colonial life dried tobacco was quite valuable and often was exchanged directly as currency. Seeds were precious. The increasing taxes placed by the British on tobacco and other imported and exported goods was one significant factor in starting the American Revolution. Today, the main tobacco-growing region in the United States ranges from Georgia to Maryland. Fertilizers and other soil management techniques, such as crop rotation, have diminished the significant soil-depletion characteristics of the tobacco plant. The World Bank no longer supports tobacco cultivation for the economic development of nations.

2.1.1. Tobacco Use

Dried and chopped tobacco was traditionally most often smoked in dry or water *pipes* and is still used this way by some in the Western and many in the Eastern Hemispheres. *Cigars*, which are tightly wound tobacco leaves, were common but now seem to be rarely used in U.S. public places—most likely due to their very strong, distinctive, and often irritating odor. A significant recent exception to this observation seems to be certain meeting rooms, bars, and lounges that cater to cigar smokers. *Snuff* is finely ground tobacco that is inhaled but is now rarely used in the United States. *Chewing tobacco*, also known as "smokeless tobacco," is placed in the mouth; as neither snuff nor smokeless tobacco is burned, environmental tobacco smoke is not a concern with their use. In pre-Columbian America tobacco was brewed or steeped like tea. The resulting strong liquid was drunk or inhaled, but this was a very dangerous practice (Gately 2001).

The most frequent use of tobacco in the United States is now via *cigarettes*. Cigarettes are often much more than just chopped, cured tobacco and paper wrappers; many brands are highly modified with hundreds of trace ingredients and chemicals added for flavoring or to control the burning rate. They are sold and "smoked" annually by the billions, thus delivering nicotine and combustion products to consumers. Most commercially produced cigarettes include some type of filter; homemade cigarettes often do not. Many pipes allow the use of small filters.

2.1.2. Equivalent Cigarettes

Because cigarettes are the most widely smoked-in-public tobacco product in the United States, those studying or designing for ETS often consider their use exclusively or convert other uses of tobacco to approximate numbers of equivalent cigarettes. As cigarettes vary greatly in their size, weight, and composition, and smokers use them in somewhat different ways, stating a value as "equal to two cigarettes," for example, implies some averaging. Trying to do repeatable studies of airborne ETS for similar rooms or occupancies is difficult. For experimental research projects, standardized smoking machines and cigarettes that can produce consistent concentrations of ETS are available.

2.2. HEALTH CONSEQUENCES

Tobacco use and ETS have been linked to various and often severe or fatal health problems (e.g., NIH 1999; IARC 2002). Many studies have shown that exposure to tobacco causes serious health consequences, in both adults and children, including *asthma*, *bronchitis*, *cancer in several organ systems*, *heart disease*, *strokes*, and *reproductive problems*, for example. Tobacco use, direct and indirect, is the leading cause of death in the United States, claiming about 440,000 Americans each year (McGinnis and Foege 1993; DHHS 2002; Mokdad et al. 2004; U.S.S.G. 2004); indirect use alone may claim many (CDC 2002; Vineis 2005, IARC 2002; Brennan et al. 2004). The World Health Organization (WHO) has estimated that almost five million people per year die worldwide due to tobacco use (WHO 2002). A few studies argued that low use of tobacco, or airborne concentrations of ETS, have little effect on health (e.g., Enstrom and Kabat 2003); these studies have received considerable criticism, and other studies refute their findings (e.g., Steenland et al. 1998; Whincup et al. 2004). There is now consensus that nicotine, delivered by tobacco use, is addictive. Users, at least initially, choose to consume tobacco products, but most often this choice is made in adolescence under the influence of advertising, peer pressure, and other factors. Tobacco smoking by adults is legal in the United States, but the prevalence is decreasing with growing restrictions on indoor smoking, increasing taxation, and more effective education. Smoking is being stigmatized too, to some extent, in our society. Tobacco users typically pay higher premiums for health and life insurance or may be denied coverage outright. Smokers often have shorter life spans, about 10 years on average (Doll et al. 2004). Most smokers, when asked, will state that they would like to quit smoking.

In our professional HVAC design work, we do not address public policy; however, it is logical that reducing or eliminating airborne concentrations of ETS and other contaminants indoors is better than having higher levels present. A zero concentration of ETS is theoretically best; health authorities argue that even extremely low concentrations of a regulated agent may have adverse consequences, and risks climb with multiple exposures (Whincup et al. 2004); both workplace and residential ETS have been implicated in its adverse health consequences. There is strong evidence that making smoking more difficult, such as via bans on smoking in indoor public spaces, leads to improved health in the population (CDC 2000). But the question of *acceptable levels* of ETS for healthy workplaces, and whether any exist, are matters of current research and debate. Cognizant health authorities have not found a safe level of exposure. Odors and irritations due to contaminants such as ETS, are, however, somewhat controllable via our decisions and actions as HVAC designers or building users.

2.3. SMOKE PRODUCTION

When tobacco or another organic material "burns" in air, the substance is being *oxidized* through *combustion*. A *flame* is rather rapid oxidization, but most tobacco is typically burned at the much slower oxidation rate called *smoldering*. The efficiency of combustion is normally inversely proportional to the rate of oxidation, so in theory smoldering should produce more complete combustion than a flaming "fire" (Kuo 1986). But in practice, low-temperature smoldering of tobacco in air produces very incomplete combustion (e.g., SFPE 1995), and thus many chemical species are produced.

When methane (CH_4) is burned in air, with complete combustion and no excess air, only carbon dioxide (CO_2) and water (H_2O), normally in its gaseous form "water vapor," are produced. However, as more complicated substances and mixtures are burned, many more complex combustion products, and often incomplete combustion, result. Tobacco, with any added enhancements, is a complex material. Table 2.1 shows that, for a typical cigarette, compounds other than carbon dioxide and water result (ASHRAE 1999, ch. 44). Many hundreds or thousands of others have

Table 2.1. Individual Combustion Products in Tobacco Smoke[1]

Major Gaseous Contaminants in Typical Cigarette Smoke

Contaminant	Weighted Mean ETS Generation Rate, mg/cigarette	Weighted Standard Error, mg/cigarette	Method
Oxidants			
NO_x	1.801	0.032	Continuous chemi-
NO	1.647	0.032	luminesent analyzer
NO_2	0.198	0.005	
Reducing Agents			
Carbon monoxide	55.101	1.064	Nondispersive IR
Nitrogen Compounds			
Ammonia	4.148	0.107	Cation exchange cartridge
Aldehydes			
Acetaldehyde	2.50	0.054	DNPH cartridge
Formaldehyde	1.33	0.034	
Ketones			
Acetone	1.229	0.040	DNPH cartridge
Nitrogen Compounds			
Acetonitrile	1.145	0.046	Sorbent tube/GC
Aromatic hydrocarbons			
Benzene	0.280	0.005	Sorbent tube/GC
Toluene	0.498	0.011	Sorbent tube/GC
Xylenes	0.297	0.006	Sorbent tube/GC
Styrene	0.094	0.002	Sorbent tube/GC
Sum of five other aromatics	0.140	0.005	Sorbent tube/GC
Alkenes			
Isoprene	6.158	0.123	Sorbent tube/GC
1,3-Butadiene	0.372	0.013	Sorbent tube/GC
Terpenes			
Limonene	0.261	0.005	Sorbent tube/GC
Heterocyclics			
Nicotine	1.585	0.042	Sorbent tube/GC
Pyridine	0.218	0.005	Sorbent tube/GC
Substituted pyridines	0.569	0.013	Sorbent tube/GC
Summary VOC Measurements			
Total HC by FID	27.810	0.483	FID
Ttl sorbed & ID'd VOC	11.270	0.525	Sorbent tube/GC
Total sorbed VOC	19.071		Sorbent tube/GC
Particles			
Respirable	13.674	0.411	Gravimetric

DNPH = 2,4-dinitrophenylhydrazine; GC = gas chromatography.
Source: Martin et al. 1997.

1. From ASHRAE 1999, ch. 44.5.

been measured as well (e.g., Table 2.1 of NIH 1999). Some additional trace compounds, below current detection technology, are likely to be present too.

Tobacco smoke that is first *inhaled* and then *exhaled* before testing has a composition different from smoke not inhaled. Human lungs are fairly good filters and absorbers. People are also sources of additional water vapor, carbon dioxide, and trace compounds; some of these compounds are air contaminants too and are known as *bioeffluents*. Dilution and removal of these contaminants from the various sources, replenishment of oxygen, and sometimes thermal control are why we ventilate buildings. There are potentially thousands of individual combustion products in tobacco smoke; only some are shown in Table 2.1.

2.3.1. Residuals

Solids are also produced when tobacco is smoked. Char, ash, and "butts," some still smoldering, must be collected for sanitary and safety reasons, at both in- and outdoor stations. Various designs of *ashtrays* and *urns* are available, and the architect and/or building owner should select and place these receptacles with care. You, as the HVAC designer, should give others some guidance on their placement relative to the ventilation systems— for example, advise that outdoor stations should not be near air intakes, as shown in Figure 2.1.

2.3.2. Smoke Flows

Figure 2.2 shows that for a cigarette, or similarly for a cigar or pipe, the smoke produced is described as *mainstream* and *sidestream smoke* (Jadud and Rock 1993). When a smoker inhales through a lit cigarette, the cigarette smolders more rapidly due to the increased availability of oxygen. The resulting combustion products, after any incorporated filtering, are drawn directly into the user's respiratory system; this smoke flow is called the mainstream.

As an ignited tobacco product awaits the next use, the smoldering tobacco still produces gaseous combustion products but typically at a much slower rate. These standby combustion products are known as sidestream smoke. They are emitted to the surrounding air from various parts of the product but mostly from the burning tip. The vast majority of this

Figure 2.1. Placement of an air intake over a small building's main entry door. Due to the location of the urn and smokers, outdoor ETS has often been entrained and distributed throughout the building (Rock and Moylan 1998).

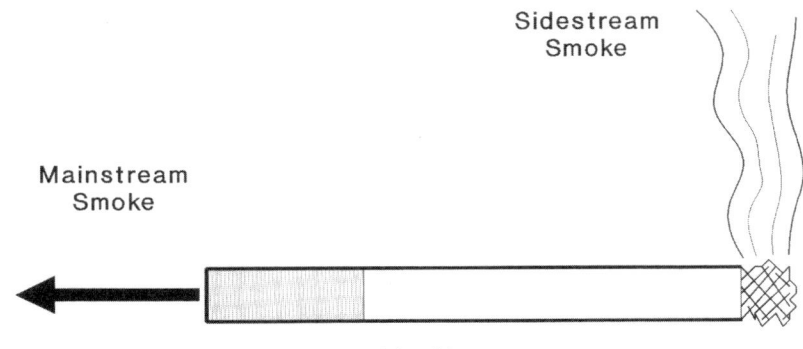

Figure 2.2. Mainstream smoke is smoke intentionally inhaled by the user; sidestream smoke is unused by-products.

sidestream smoke, which is produced at a lower temperature and thus has somewhat different products, has not passed through any filter that may be present in the smoking device nor has it been inhaled by the smoker. Thus, it normally has higher concentrations of combustion products than exhaled mainstream smoke. Nonsmokers inhale diluted sidestream smoke as well as exhaled mainstream smoke.

2.3.3. Smoke Exposures

Related terminology is used to describe individuals' exposures to airborne tobacco combustion products. Figure 2.3 shows that the intended user's mainstream smoke is called *firsthand* or, more commonly, *active* or *direct* exposure. Breathing the sidestream and/or exhaled mainstream smoke is *secondhand* or *passive* exposure. The user of that particular lit cigarette, cigar, or pipe receives both direct and secondhand smoke exposures, while nearby people, both smokers and nonsmokers, may be receiving significant secondhand smoke exposures. The degree of secondhand smoke exposure is highly dependent on the concentration of ETS present at the particular location, the time residing in that environment, and the respiration rate. Smoke may also cause irritation of the eyes and/or skin, so considering only "the breathing zone" is not likely sufficient when designing for ETS or other irritants. Smoke can travel significant distances and still be objectionable. Also, its toxins may be present without noticeable irritation reactions. Some people are irritated by even low exposures to secondhand smoke, while others, especially many smokers, are much less sensitive. ETS's irritations versus exposures is a topic of current study by others.

2.3.4. ETS

Environmental tobacco smoke, or ETS, is a term or abbreviation often used in a broad sense to mean tobacco smoke that is present indoors or out, but it is usually smoke that is airborne within a space or building. ETS is secondhand smoke.

While nothing can be done from an HVAC point of view to reduce a smoker's active exposure to his or her own tobacco product, good design, construction, operation, and maintenance of ventilation systems can significantly reduce indoor secondhand smoke concentrations. While poten-

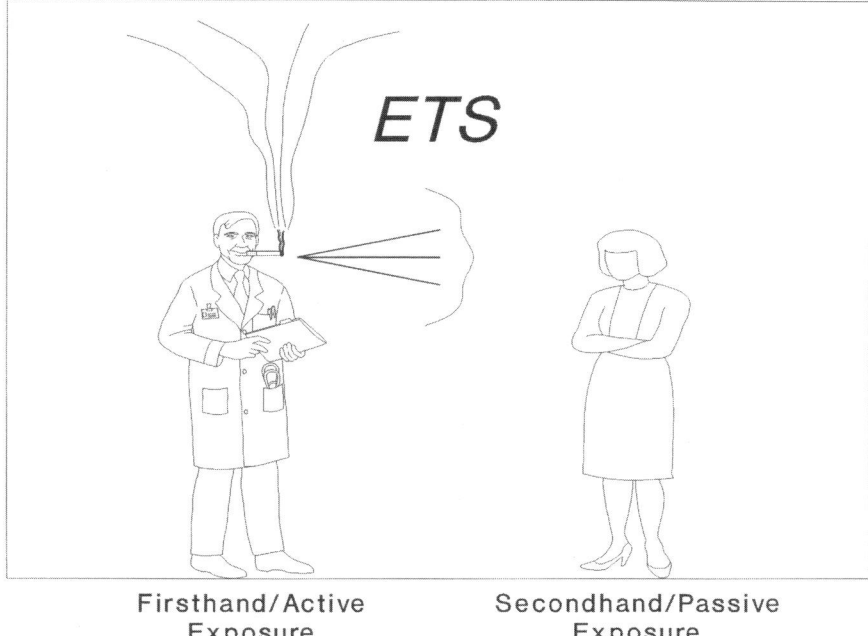

Firsthand/Active
Exposure

Secondhand/Passive
Exposure

Figure 2.3. An "active" smoker receives firsthand or direct dosage while others are "passively" exposed to secondhand smoke.

tial reductions in any health risks or liabilities cannot be claimed through use of the information provided in this book, hopefully improved ventilation can reduce odor and irritation complaints.

2.3.5. Where Does the Smoke Go?

While each compound in tobacco behaves differently, for simplicity an overall mass-balance is useful in describing where the mainstream, sidestream, and exhaled smoke ultimately goes. Figure 2.4 shows that for active exposure many combustion products are *absorbed* into the body via the trachea and lungs. Some may condense in the mucous and be swallowed or expectorated, and a large portion of the inhaled smoke is expelled via respiration. These processes are highly transient, and many experimental and analytical models of such have been produced for purposes such as predicting exposures.

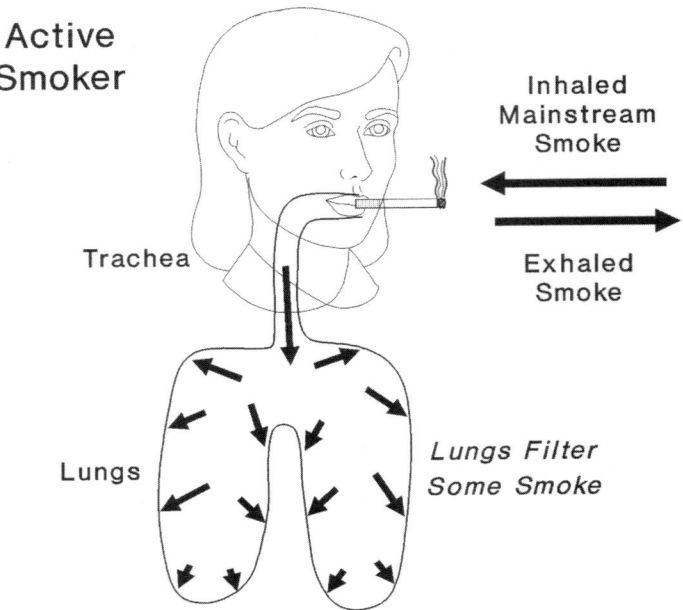

Figure 2.4. When smoke is inhaled, a portion of it is filtered and some is absorbed; the exhaled smoke is a composition different from that inhaled.

Sidestream and exhaled smoke enters the surrounding air and mixes with it to varying degrees in ways that will be discussed in later chapters. As shown in Figure 2.5, this ETS is stored in the room air; expelled with the exhausted "old" room air; absorbed into and/or *adsorbed* onto surfaces, furnishings, plants, and clothing; and is also inhaled by occupants, all to varying degrees. Some of the inhaled ETS is exhaled and then inhaled again by that person or others. The uses of rooms, the introduction of air, the current presence or not of occupants, and all the ETS storage and removal processes are highly transient, so it is difficult to accurately describe the ultimate fate of a particular ETS component. Instead, in our design work we must often consider:

1. How many smokers are likely to be present.

2. How many total occupants are likely to be in the space.

ETS is adsorbed onto surfaces
and absorbed into materials

Figure 2.5. Airborne smoke is diluted by the room air, and much is carried away via exhaust or exfiltration. Some ETS sticks to surfaces or penetrates materials and later may become airborne again.

3. What the smoking rate is.

4. What the needed air-flow rates and ventilation equipment are.

The transient effects are largely ignored, except that we should normally continue ventilating for a period after the expected occupancy ceases to help decrease residual ETS concentrations.

2.4. AIRBORNE POLLUTANTS

In general, airborne pollutants of common concern in buildings, including ETS, are divided into two broad classifications: *particulates* and *gases*, with *volatile organic compounds* (*VOCs*) being one class of gases of specific interest (Rock and Zhu 2002). Particulates vary in size from, for example, large dust accumulations that are not easily inhaled, to microscopic specks, such as pollen, that are easily drawn into the lungs and deposited there, transferred to the bloodstream, or exhaled. The variously sized particulates may be either liquid or solids. Small airborne solid particles or liquid droplets are often called *aerosols*. *Respirable suspended particles* (*RSPs*) are small, easily made airborne particles and can be actively measured in near real time with appropriate sensing equipment.

2.4.1. Odors and Irritations

Odors are perceptions through our nasal receptors of airborne materials—either particles or gases. Odors or "smells" range from very pleasant, such as that of some perfumes and great foods, to neutral, to the highly offensive, such as sewage and ammonia fumes. Some materials such as "pure" air are perceived as odorless even though the components—N_2, O_2, and so on—are present in high concentrations. Others in very low concentrations give a moderate to strong sensory perception. Users of tobacco often rate the quality of the products by their odors and tastes. Nonsmokers, and smokers too, find various levels of ETS as imperceptible, acceptable, or intolerable (ASHRAE 1991).

Irritations can be physical, such as watering eyes, runny nose, skin rashes, or the scratching response, or psychological. Some sources of irritants are not airborne, such as residual cleaning chemicals left on surfaces, and thus we can't control them well with HVAC. But our ventilation systems and other measures can often reduce the levels of airborne irritants. However, when a person has an irritation response to any tobacco use, our efforts may not be successful. Removal of the smokers from the space may be the only effective action to resolve complaints; such measures are generally not under the control of HVAC designers. Some airborne irritants can have adverse health effects and may already be regulated by OSHA, for example. The effects of acrolein, a colorless poison and irritant and a component of ETS, are the subject of ongoing studies (e.g., Nazaroff and Singer 2002).

2.4.2. Sources and Sinks

There are many *sources* of airborne pollutants, odors, and irritants in and around buildings; one particularly potent source, if present, is tobacco smoking. Within buildings, other sources are the occupants themselves, through respiration, perspiration, and their activities. Additionally, building materials, furnishings, and consumables off-gas, and molds may grow in quantity if conditions are suitable. Removing pollutant sources from buildings, such as tobacco use, or providing local exhaust for the pollutants they generate, can be very effective. But such measures are often not appropriate or feasible, so dilution or replacement of the "dirty" or "used" air with cleaner air is more common. As previously mentioned, there are also pollutant *sinks* within buildings that remove airborne contaminants, including those from ETS. Particulates will settle onto or adhere to materials, and VOCs can be absorbed by surfaces at various rates, for example. Some pollutants are reemitted later so sinks can also become sources of contaminants. Air cleaners, discussed in some detail later, are intentionally enhanced sinks.

2.4.3. Sensing and Control

Concentrations of gaseous contaminants, such as some of those present in ETS, can be measured, so it may be possible to automatically control our engineering measures to suit the conditions at any given moment. More reliable and economical ETS sensors are needed, however, for more widespread use. But a common problem is defining at *which* concentrations, for *how long* exposures, and for *which* particulates and gases the indoor air quality is deemed "acceptable." Designing for environmental tobacco smoke is frequently of concern and is especially challenging because it contains both particulates and gases, and no "safe" concentrations of and exposure times to ETS have been established by the relevant regulatory or health agencies. In theory, a zero concentration for an indefinite time period would be the baseline for healthful air. At the other extreme, a tobacco smoke–filled room with poor ventilation has high concentrations of ETS's various components, many of which are known carcinogens and irritants. The combined effect of the potentially thousands of chemical species in secondhand smoke, even at separate acceptable health risk levels, is not known (ASHRAE 2003c).

In practice, even the outdoor air has measurable quantities of many contaminants present in ETS, from various human and nonhuman "natural" sources (Nelson et al. 1998). If selecting sensors for measuring specific components of ETS, be aware that "background" levels may be greater than zero.

As described in this and later chapters, contaminant concentrations, and thus *indoor air quality (IAQ)*, are significantly transient, so measures meant to enhance the indoor environment should either seek to address the near-worst expected conditions or adapt to meet varying circumstances. Diluting with cleaner air, exhausting contaminants, filtering any reused air, and source controls are widely used methods for providing acceptable IAQ in many commercial and institutional buildings, with or without ETS present. These and other design approaches will be discussed further in the following chapters.

3

INDOOR ENVIRONMENTAL QUALITY

This chapter reviews some HVAC concepts that are also essential for successful ETS design. If unfamiliar with one or more of the topics, you are encouraged to seek more information via the mentioned references.

Indoor environmental quality (*IEQ*) is a term that describes the total indoor experience. It includes the acceptability to a particular person of the temperature, humidity, air movement, odors, irritants, contaminants, sounds, and vibrations. But it also includes other factors that are well beyond the control of HVAC practitioners, such as lighting, colors, textures, furnishings, job satisfaction, and personal relationships.

3.1. THERMAL COMFORT

Thermal comfort is a basic part of IEQ. Human thermal comfort is the "condition of the mind that expresses satisfaction with the thermal environment" (ANSI/ASHRAE 2004c). The temperature of the air and surrounding surfaces, humidity, air movement, and skin moisture all affect the perception of thermal comfort. Each person decides whether he or she is comfortable or not at any particular moment, whether indoors or out. The modern descriptions of indoor thermal comfort, and the testing methodologies for such, can be found in the cornerstone book *Thermal Comfort* (Fanger 1972).

3.1.1. ANSI/ASHRAE Standard 55

As individuals' perceptions of thermal comfort vary, it is not possible to satisfy all occupants at any particular moment. A historic goal was 80% acceptability, but in recent years 90% satisfaction is often the objective. *ANSI/ASHRAE Standard 55, Thermal Environmental Conditions for Human Occupancy*, describes the current consensus of what's needed to maintain thermal comfort for the vast majority of building occupants (ANSI/ASHRAE 2004c). Because thermal comfort is dependent on clothing, activity level, air movement, and mean radiant temperature, care must be taken when evaluating the Standard's base temperature and humidity recommendations. These recommendations are given via the summer and winter "comfort zones," or ranges of "operative temperatures" and humidities, that produce 90% satisfaction when the occupants have a light activity level. Fortunately, many spaces that potentially include ETS have occupancies that can be described by this "light work" description of human metabolic rate. But occupants in some spaces with ETS, such as those in dance halls and manufacturing facilities, can have significantly higher activity levels, and thus the parameters such as temperature, relative humidity, and air velocity must be adjusted through the methods presented in Standard 55 and elsewhere to ensure thermal comfort for most occupants of these spaces.

The comfort zones in Standard 55 are shown on abridged psychrometric charts. For an introduction to *psychrometrics*, which is the study of moist air, see the *ASHRAE Handbook* (ASHRAE 2001, ch. 6) and ASHRAE's *Psychrometrics: Theory and Practice* (ASHRAE 1996). Visit the ASHRAE Bookstore, via www.ashrae.org, to obtain Standard 55 and the various other ASHRAE publications mentioned in this book and elsewhere.

3.1.1.1. Comfort Conditions and ETS

Cain et al. (1983, 1984) found a strong correlation between thermal comfort and humans' reactions to ETS's odors and irritants. At low to mid temperatures and humidity, sensitivity was relatively constant. However, at higher temperatures and/or relative humidities, sensitivity to ETS and other odors grew rapidly and thus higher ventilation rates would be needed for acceptability. Therefore, it is critical that smoking-optional spaces be conditioned to maintain thermal comfort, and the suggestions that follow in this book assume such is done.

3.1.2. Thermal Zoning Is Important

Thermal zoning is the art of grouping spaces together that have similar thermal characteristics. Only one HVAC system or subsystem then typically serves a particular zone, rather than having a separate unit for each space. Thermal zoning is thus a technique intended to reduce the *initial* or *first cost* of HVAC systems, but thermal comfort can suffer with poor zoning decisions. These zoning decisions are normally made before performing the HVAC load calculations.

In designing for ETS, it is desirable that spaces with ETS be in separate thermal zones from non-ETS spaces. A significant exception is when all the ETS-laden air in the smoking-allowed portion of a larger, multiroom zone is exhausted directly to the outside. This identification of ETS spaces, their purposes, the expectations, and the various HVAC systems that serve the spaces is a significant part of the rest of this book.

3.1.3. Acoustics

While not part of thermal comfort, the acoustical environment does affect overall comfort and thus the perception or not of good overall IEQ. *Sound*, which is rapid pressure variations in gases or liquids, is a complex mode of energy transfer, and the human perception of sound levels is highly nonlinear. *Noise* is sound that is perceived to be objectionable but can be described anywhere from "quiet" to "loud," or even truly painful and damaging. *Vibrations*, whose physics are very similar to that of sound, are energy transfers via the rapid, repeated physical movements of solids such as buildings' structural elements. Mechanical equipment is often the source of objectionable sounds and vibrations; in ETS-specific systems exhaust fans and any makeup air units' fans and compressors are possible sources of undesirable noise and/or vibrations. But beneficial sounds can also be produced by HVAC systems; for example, air diffusers often produce a sound that is similar to "white noise," and this sound can be useful at reasonable levels for "masking" undesirable noises or conversations.

While air-flow issues dominate the discussion of ETS, problems with thermal comfort, noise, and vibrations can easily cause an occupant's perception of poor IEQ. Thus, these factors need to be addressed by the HVAC designer, the architect, and others too, or it is possible that otherwise appropriate designs for ETS may fail to please. More detailed information on thermal comfort and acoustics can be found in Rock and Zhu (2002) and the *ASHRAE Handbook*, for example.

3.2. INDOOR AIR QUALITY

HVAC designers, in addition to thermal comfort, often focus on *indoor air quality* (*IAQ*). While providing "pure air," devoid of ETS and all other contaminants, may seem like the goal, it is unrealistic, so instead we try to achieve time-averaged *acceptable indoor-air quality* for building occupants (ANSI/ASHRAE 2001). Acceptable IAQ is needed to maintain healthy and productive indoor environments. Creating acceptable indoor-air quality involves many factors, but delivering outside air to spaces, controlling moisture and contaminants, and treating any reused air are the most common ways of doing so. For most commercial applications, the HVAC systems are expected to create and maintain these acceptable indoor-air conditions. The presence of ETS causes great concern from a design point of view, because no U.S. regulatory authority has set limits below which health effects are inconsequential or "acceptable." There are other airborne materials for which acceptable values have not been established, too. This book, therefore, can only address ETS odor and irritation control.

3.2.1. Air and Airflows

Air, both inside and outside of a building, is a mixture of gases, water vapor, and contaminants. *Outside* or *outdoor air* (*OA*) may or may not be of acceptable quality and thus may require no, little, or considerable treatment before use toward creating acceptable indoor air. ANSI/ASHRAE Standard 62.1-2004, to at least a minimal extent, requires some observation and documentation of the existing outdoor-air quality if the air is to be used for ventilation purposes; air cleaning is required in some cases. Figure 3.1 shows an *air-handling unit* (*AHU*) admitting out-

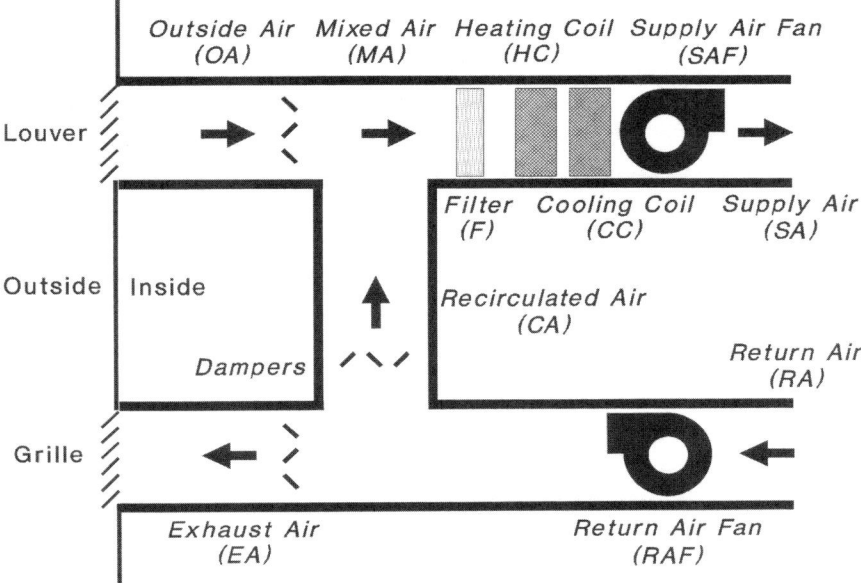

Figure 3.1. The airflows and equipment in a simple commercial, institutional, or industrial air-handling unit (Rock and Zhu 2002).

side air via a dedicated *OA* louver and duct inlet. *Return air* (*RA*) is air removed from rooms and intended for full or partial reuse. The portion to be reused is the *recirculated air* (*CA*), and the remainder is expelled from the building as *exhaust air* (*EA*). If 100% of the room air is to be directly exhausted, it is often called *relief air* (*LA*), especially when its replacement, conditioned outside air, known as *makeup air* (*KA*), causes the space to be positively pressurized relative to its surroundings (ASHRAE 2001, ch. 26).

As also shown in Figure 3.1 for a simple AHU, the recirculated air is combined with the outside air to form *mixed air* (*MA*). To minimize the number of air-treatment locations, this mixed air typically is conditioned rather than treating the upstream *OA* and *CA* flows separately, but exceptions do occur. Air treatment will be discussed in greater detail in later sections of this book. After treatment and/or conditioning, the mixed air becomes *supply air* (*SA*), which is then directly or indirectly delivered to the thermal zones and their associated rooms.

3.2.1.1. Systems and Equipment Terminology

An *air distribution system* describes all the air-flow equipment: the air handlers and their components, ductwork, dampers, terminal boxes, supply and return air terminals, and exterior intake louvers and exhaust grilles. There are various combinations of equipment that define the many different types of air distribution systems. The following are some of the more common types used in commercial, institutional, and industrial buildings.

In a *single-duct system*, only one stream of supply air is delivered via the air ducts to the connected rooms and typically provides only hot or cold air depending on the season, climate, and building operation. *Fans* or "blowers" are used to overcome air pressure losses due to "friction" in the various system components and ductwork.

Dual-duct systems, while a little more complicated, provide two streams of supply air, typically one hot and one cold, that can be mixed in various proportions to meet the thermal needs of each particular zone. To reduce energy consumption, one of the two air streams' heat exchangers (*HXs*), also known as heating or cooling "*coils*," or hot or cold "*decks*," is usually deactivated in each extreme season so that untempered mixed air is used as one air stream, rather than having both the heating and cooling airflows "fight" each other.

Unitary equipment, and in-room air handlers such as *unit ventilators*, *packaged-terminal air conditioners* (*PTACs*), and *packaged-terminal heat pumps* (*PTHPs*) are commonly used in elementary schools, motels, and elsewhere; often do not have any ductwork; and thus discharge conditioned air directly into the spaces that they serve. These unducted machines usually are placed through or near exterior walls for easy access to outside air. But they often have very limited *OA* and *EA* capacity, and thus applications such as for ETS may need ancillary makeup and exhaust air.

Rooftop units (*RTUs*) are a very popular type of unitary equipment; most admit outside air and do so through intake "hoods." Some have fan-powered provisions for exhausting air, but many have only supply air fans. Care is needed in selecting predesigned RTUs for ETS applications because they often have very limited *OA* and humidity-control capacities. But *makeup air units* (*MAUs*), which are often RTUs that are designed for 100% *OA* flow, recirculate no air. RTUs and MAUs usually are connected to ductwork, but in some cases they inject air directly into spaces. The return air, if any, to these units may be ducted or unducted.

3.2.1.2. Primary/Secondary Systems

Systems that "meter" the outside air, and often partially or fully condition it separately from the recirculated air, are becoming more popular with designers and owners/operators. They can provide some more assurance that the desired quantities of outside air are being introduced to the buildings, but they are typically more complex and of somewhat higher initial cost than simple systems. Figure 3.2 shows the supply-side of one of these two-level systems, known as *primary-secondary HVAC systems* (ASHRAE 2004, ch. 2).

The primary air handler is the one that admits the outside air. It may or may not allow recirculation of some or all of the return air if such air is gathered and brought back from the zones instead of being exhausted directly from the spaces. With primary/secondary systems, the supply air from the primary air handler is renamed *primary air* (*PA*), and this air is

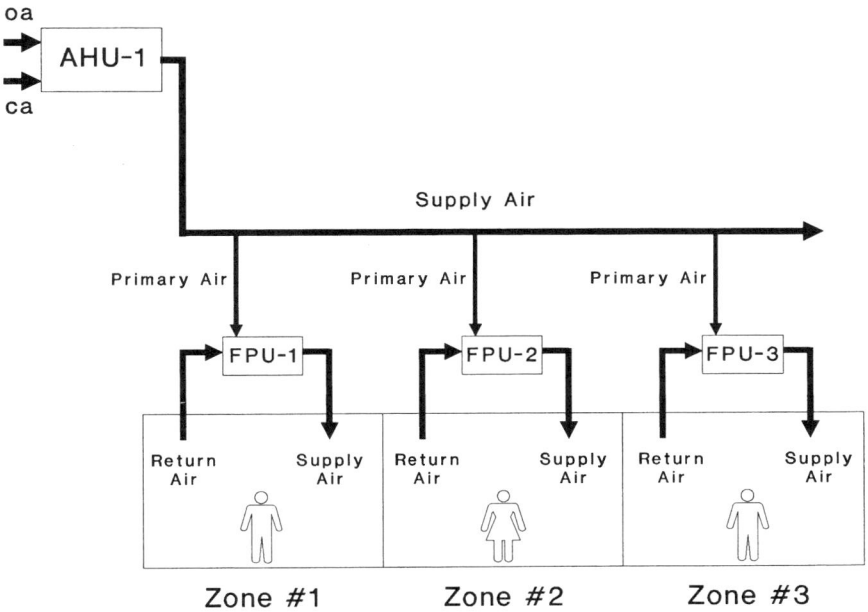

Figure 3.2. When using two levels of air handlers in series, the supply air is renamed "primary air" as it reaches the secondary "terminal units." This primary air is often used to deliver ventilation air to the zones (Rock and Zhu 2002).

sent to the secondary-level air handlers. These secondary units are normally placed within each zone—often in their ceiling or floor *plenums*, or over adjacent hallways for noise control. The secondary air handlers often mix the primary air with air recirculated from the zones. These local AHUs and *terminal units*, often called "mixing boxes" or "fan coils," may filter and/or condition the combined air, and their resulting air flows become the supply air streams that are ultimately delivered to the spaces (ASHRAE 2004, ch. 17). When a primary/secondary system employs heat recovery in the primary air handler, it is called a *dedicated outdoor air system* (*DOAS*). Normally the goal of the DOAS primary air handler is to "pick up" all of the ventilation load while the secondary units handle the room loads (Mumma and Shank 2001).

3.2.1.3. Readmission of Exhaust Air

Due to air mixing outdoors, some quantity of exhaust from various sources will be admitted to buildings. When exterior conditions are unfavorable for good dilution and dispersion, or possibly due to poor design or later construction, much more exhaust or relief air may be drawn into a particular air intake, as shown in Figure 3.3. This *reentrainment* of "used" air, also known as *reentry*, is undesirable when substantial but is inevitable (ASHRAE 2003b, ch. 44; ANSI/ASHRAE 2001). Exhausts carrying ETS should be designed to minimize reentry, and, if the flows are significant, these ETS-bearing exhausts can be designed like vents from kitchens, restrooms, or possibly chemical fume hoods. More guidance on this topic can be found later in this book, in codes (e.g., the Uniform Mechanical Code [UMC]), and Chapter 44 of the *ASHRAE Handbook* (ASHRAE 2003b). Admission of fumes from other sources, such as from plumbing vents or kitchen exhausts, should be minimized too.

When complaints are received, or anticipated, about readmission of ETS-laden exhaust air, various corrective measures are available—for example, increasing discharge velocities, stack heights, directions, locations, power-dilution with outside air, and so on—and can usually help. However, in a densely built location with often reduced wind velocities— for example, in the city-center of a major metropolis—it may become necessary to clean the exhaust air to some degree before discharging it in order to reduce complaints from neighbors. This treatment can range from minimal filtration to more complete "scrubbing" via chemical and/or

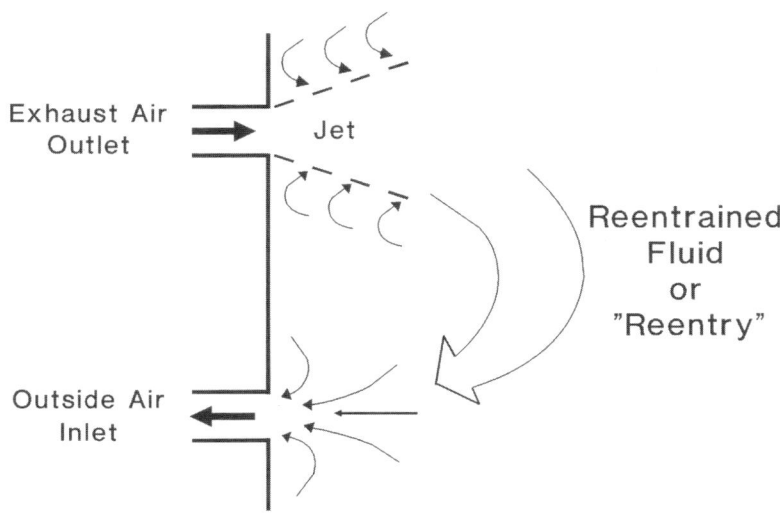

Figure 3.3. Due to the nature of jet flows and inlets, there will almost always be some degree of "reentry" of exhaust. A goal is to minimize this reentry to an acceptable "dilution ratio" (Rock and Moylan 1998).

electromechanical processes such as those used in the power-generation industry (El-Wakil 1984, ch. 17), but on a much smaller scale. There may be market potential for more manufacturers to develop economical, small-scale ETS scrubbing systems.

3.2.2. Indoor/Outdoor Air Exchange

Air exchange, between the indoors and outdoors, can be accomplished in various ways, as Figure 3.4 shows. *Ventilation*, the intentional introduction of outside air into a building, can be achieved via two approaches, each with its own advantages and disadvantages, and a combination is possible. *Natural ventilation*, which allows air to move through a building via intentional openings, is driven by wind and buoyancy forces. Natural ventilation requires no energy to drive a fan, but it is highly dependent on the current weather conditions and often on active occupant participation. As such, natural ventilation is not often a good choice for ETS design.

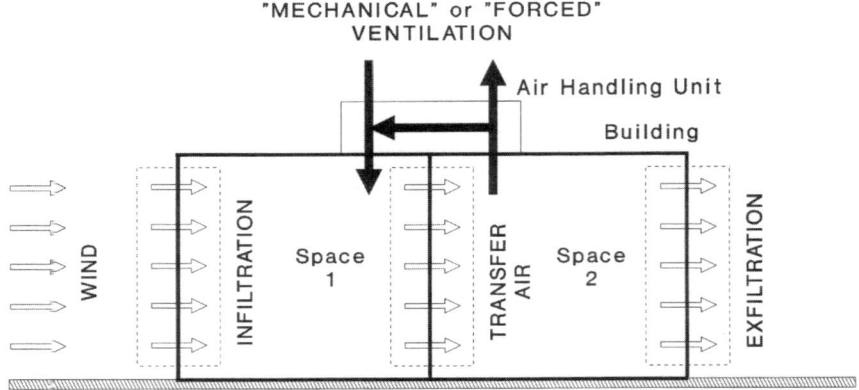

Section View

Figure 3.4. Air exchange, of outside air for indoor air, takes place via multiple routes in modern buildings (Rock and Zhu 2002).

Exceptions might be semiindoor spaces such as covered entrances, patios, and other "outdoor" spaces where smoking is allowed and natural air currents are substantial.

Mechanical or *forced ventilation*, which requires an external driving mechanism such as a fan, is predictable, controllable, and normally highly reliable, but energy, equipment, and maintenance costs are incurred. For ETS design, mechanical ventilation, as was already described in this chapter and will be further explored in this book, is the logical choice for indoor spaces.

Infiltration, which is the unintentional "leakage" of unconditioned outside air into a building, is uncontrolled. For exterior zones, where infiltration may cause thermal discomfort, slight positive air pressure differentials are often created by HVAC designers to minimize air leakage into buildings. A smoking room, with a negative pressure relative to the outside, may have significant infiltration, however. No credit is usually taken for infiltration when designing for ETS due to infiltration's variable nature. *Exfiltration*, which is the unintentional and uncontrolled leakage of air out of the building, is usually not considered in ETS design due to the unpredictable flow rates and paths.

As outdoor air, either via ventilation or infiltration, is often at temperatures and humidity levels significantly different from that desired indoors, considerable energy and equipment are needed to condition it. A common goal in ventilation design, including for ETS, is to optimize the quantity of outside air that needs conditioning. Not only might the energy consumption be reduced and thermal comfort increased, but the physical size and initial cost of the HVAC system might decrease with such optimization.

Transfer air is the movement of indoor air between spaces and is normally accomplished via *transfer grilles* or *ducts*, *open door-* or *entryways*, *continuous ceiling* or *floor plenums*, *undercut doors*, or *transoms*. Air pressure differentials created by the HVAC system, the stack effect, infiltration/exfiltration, and any natural ventilation drive this internal air movement, but in rare cases fans between spaces are used to increase the transfer air-flow rates. Check local codes, as some methods of increasing transfer air rates, such as door-mounted transfer grilles or connected plenums, may be prohibited due to fire and smoke movement concerns. Noise also travels with transfer air. But, when possible, this transferred air has likely already been conditioned and may be usable as is, so large energy savings are possible. To some degree the transfer air has already been "used" for ventilation purposes, so care is needed in identifying its remaining usefulness. As will be discussed later in detail, transfer air from less contaminated areas is often used to ventilate ETS areas.

3.2.2.1. Ventilation Air

Ventilation air is defined in Standards 62.1, 62.2, and elsewhere as being outside, recirculated, or transfer air, or a combination, treated or not, that is intended for use in creating acceptable indoor air quality. Normally, this ventilation air requires some amount of treatment, as will be described later, but when acceptable outdoor air (ANSI/ASHRAE 2004a) is available, only minor particle filtering is required. "Ventilation air" is often confused with Section 3.2.2's term *ventilation*, probably because mechanical ventilation is often the only planned source of ventilation air for a commercial or institutional building. However, treated recirculated air and that unused portion of transfer air are often employed as part or potentially all of the ventilation air when designing for ETS and some other contaminants.

While not specifically addressed in this book, ventilation air for low-rise residential buildings in the United States is typically provided via infiltration and natural ventilation, but user-controlled exhaust fans are often employed in kitchens and bathrooms to increase these air-flow rates.

Mechanical ventilation, sometimes with incorporated air-to-air heat recovery units, is seeing increased use in residences, especially in those that are tightly constructed to minimize infiltration and exfiltration. The new ASHRAE Standard 62.2 (ASHRAE 2004b) addresses these topics further.

3.2.2.2. The Outside-Air Fraction

While recirculation of all or part of the return air from a smoking area to other spaces will not be recommended in this book, in some cases, especially with effective air cleaning, recirculation may be possible to the ETS area itself. As shown in Figure 3.1, this treated recirculated air will ultimately be reused as all or part of the supply air. In addition, the outside air is usually provided via the supply air, as is also shown in the figure. The *outside-air fraction* (X_{oa}), or if expressed as a percentage, the *percent outside air*, is how this reuse of air, and introduction of outside air as part of the supply air, is described via one numerical value. The outside-air fraction for a simple air handler is (ASHRAE 2001, ch. 26)

$$ X_{oa} = \frac{Q_{oa}}{Q_{sa}} = \frac{Q_{oa}}{Q_{ma}} = \frac{Q_{oa}}{Q_{oa} + Q_{ca}} \tag{3.1} $$

where Q, or often $\dot{V}$, as is used in the remainder of this book, is the volumetric flow rate of a particular air stream in ft^3/min (CFM) or l/s (LPS). Characterizing the needed flow rate of outside and/or treated recirculated air is one of the major goals of this book and Standard 62.1, but the total flow rate of supply air is determined via thermal loads calculations which are addressed elsewhere such as in the *ASHRAE Handbook* and in Pedersen et al. (1998).

For *constant air volume* (*CAV*) systems, where the supply air-flow rate is relatively consistent during normal hours of building operation, the outside air fraction is frequently set to a constant value, usually by one or more postconstruction *test*, *adjust*, and *balance* (*TAB*) technicians and/or *commissioning agents*. In "high-recirculation" systems, this percent of outside air is often 10% to 20% for typical commercial and institutional buildings. For spaces where ETS is present, and even with good recirculated air treatment, if any, the percent outside air may be much higher. In some cases, as with some reduced room air-flow regimes or when effective ETS treatment in the recirculated air is not possible, "once-through" or "100% OA" systems may be appropriate. *Demand controlled ventilation (DCV)* is where the quantity or percent outside air is adjusted up or down, in real time, to account for changes in occupancy levels and uses; this will be discussed in more detail later.

Variable air volume (VAV) systems are where the zones' supply air-flow rates are intentionally varied to meet the changing thermal loads of the zones. In VAV systems, the outside air fraction must also be adjusted to ensure that the required flow rate of ventilation air is achieved, unless the ventilation air needs are met at the lowest VAV flow rate (ANSI/ASHRAE 2001; Standard 62, addendum *u*). Demand controlled ventilation can also be employed with VAV systems.

An *air-side economizer* is an energy-saving control scheme for systems with normally low-to-mid percent outside air. When indoor cooling is required, and the outside air is cool, the outside air fraction is adjusted upward from the base value until the thermal loads are met. During this air-side economizer operation, much more than the needed ventilation air is typically admitted to the air handler and is a side benefit of its use. However, in hot and very humid climates air-side economizers often introduce too much moisture to buildings and are, as such, not specified. *Energy recovery*, described later in some detail, can be very effective in reducing energy use or for increasing ventilation rates without incurring as big an "energy penalty." In very dry climates, such as the high deserts of the American West, well-designed and maintained direct *evaporative cooling* of the outside air stream can also significantly reduce the energy consumption required for cooling (ASHRAE 2003b, ch. 51) and potentially increase ventilation rates at times. Indirect evaporative cooling may be possible where direct is not allowed. A detailed, hour-by-hour *energy analysis*, which is different from and in addition to the peak thermal load calculations, may be required to evaluate alternatives.

3.2.3. Room Air Diffusion

The movement and mixing of air within rooms is called *room air diffusion*, and the particular characteristics of such can help or hinder efforts to address airborne contaminants. The placement and performance of the *supply air outlets* have substantial effects on air-flow patterns produced in spaces, so these outlets or *diffusers* should be selected with great care. The locations of the *return air* and *exhaust air inlets* or *grilles* are much less important, except when placed very near heat or contaminant sources to hasten removal somewhat.

Perfect mixing, shown in Figure 3.5, is the *theoretical* condition where the air properties throughout the room, and in the return/exhaust air, are identical due to infinitely high mixing of the air within the room. If desired, well-selected, installed, and operated supply air outlets can produce near-perfect mixing conditions. But when airborne contaminants are being produced from known, fixed-location sources, mixing may be highly undesirable. Instead, the pollutants should be removed or "vented" as much as possible before they can mix with the room air.

Entrainment flow, shown in Figure 3.6, is the most common room airflow pattern. *Jets* of air from the outlets entrain room air into them, induce room air currents, and enhance mixing. Depending on several factors, a high degree of "stirring" can occur, so near-perfect mixing can be achieved. Selection and placement of the outlets and inlets is critical so that, for example, a substantial portion of the supply air doesn't "short-circuit" to the return or exhaust without adequately mixing with the room

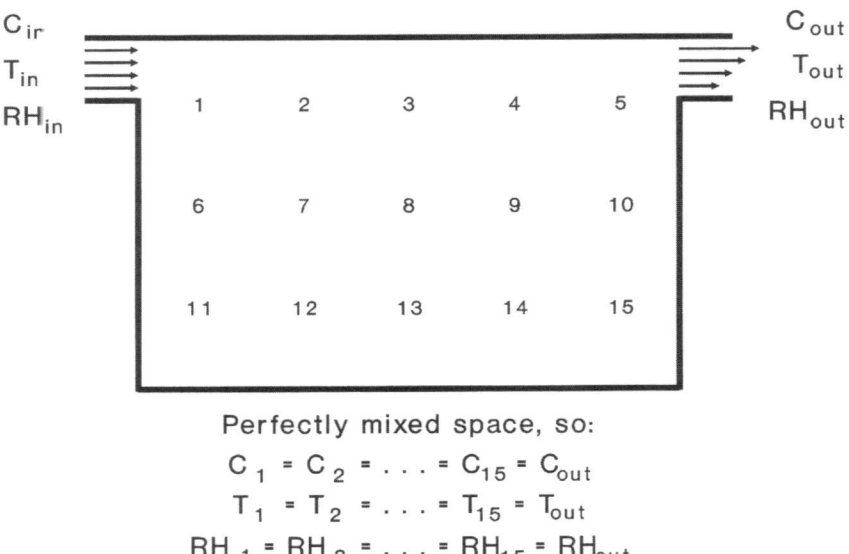

Perfectly mixed space, so:

$$C_1 = C_2 = \ldots = C_{15} = C_{out}$$
$$T_1 = T_2 = \ldots = T_{15} = T_{out}$$
$$RH_1 = RH_2 = \ldots = RH_{15} = RH_{out}$$

Figure 3.5. The air in a "well-mixed" space has properties that are consistent throughout, and the exhaust or return air thus has the same properties. However, the supply air often has a temperature different, for example, from the room air (Rock and Zhu 2002).

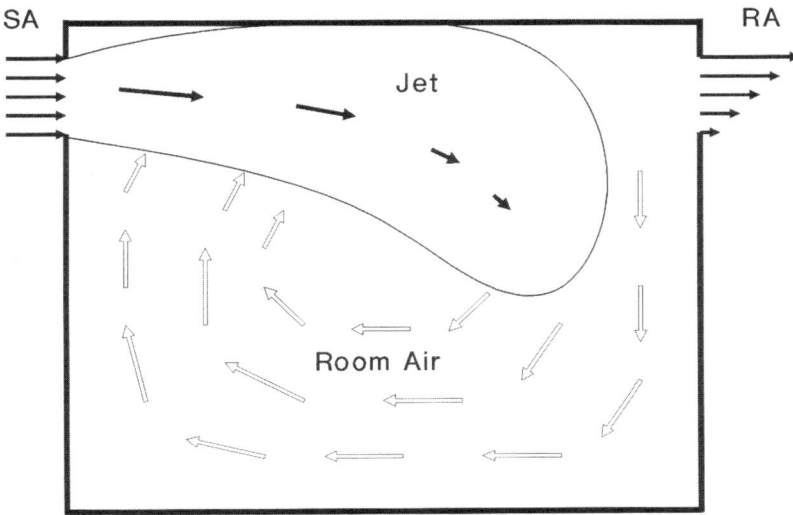

"Entrainment" or "Conventional-Mixing" Flow

Figure 3.6. Most air diffusion systems in the United States are of the conventional mixing type. When air outlets are properly selected, installed, and operated, near-perfect mixing results (Rock and Zhu 2002).

air. The designer's guide by Rock and Zhu (2002) covers the air terminals' selection process and the physical phenomena in detail, primarily for ceiling-based outlets. ASHRAE's TC 5.3, Room Air Distribution, sponsored that book's development.

Displacement flow, shown in Figure 3.7 for floor-to-ceiling flow, intentionally discourages mixing of air within the space. Instead, a "plug-" or "piston-like" air motion is desired to continuously "sweep" air through the room and is thus attractive for use in removing airborne contaminants (Chen and Glicksman 2003). While floor-to-ceiling displacement flow is becoming somewhat popular for office spaces, and its use has been standard design practice for dedicated mainframe computer rooms, other airflow directions may be preferable. For example, side-to-side displacement flow is sometimes utilized in paint booths to help keep overspray off work pieces, and ceiling-to-floor flow is common in some manufacturing "clean rooms," where heavier-than-air particles needed to be drawn quickly to the floor.

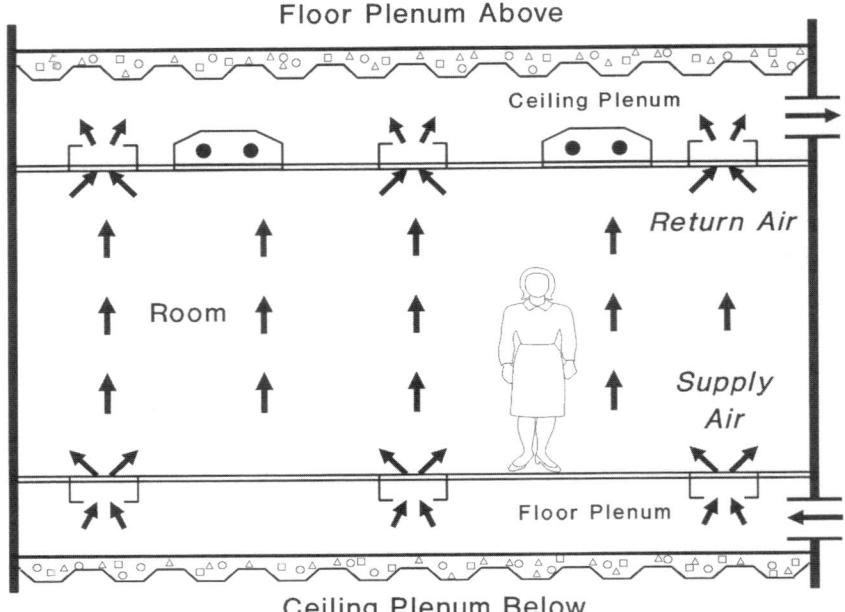

Figure 3.7. Displacement or "plug" flow intentionally reduces mixing. Underfloor air distribution, shown, encourages mixing near the feet to improve comfort but then uses low-mixing flow ("stratification") above to help condition and ventilate the space (Rock and Zhu 2002).

As will be discussed later, floor-to-ceiling displacement flow appears attractive for ETS design work because warm smoke and near-body air plumes tend to rise due to buoyancy forces (Enbom et al. 2000). However, two challenges are that true displacement flow is not possible as some mixing will occur and having the room's vertical temperature variation 5°F (3°C) or less is desired for maintaining thermal comfort (ANSI/ASHRAE 1992; Standard 55). When introducing cold air at or near the floor, discomfort can easily occur with low-activity-level occupants, so quick mixing of the cool supply air with the room air near the floor is typically encouraged to help moderate the air temperature and improve ankle-level thermal comfort for fairly sedentary occupants. This approach, which encourages low-level mixing, is often called *underfloor air distribution* (*UFAD*) (Bauman 2003), and seems attractive for ETS design where possible.

Obstructions in a room, such as furniture and decorations, often limit air movement. Sometimes regions of poor air diffusion result, but these regions often can't be predicted in the HVAC design stage of a project. Some postconstruction adjustability, if possible, is therefore desirable in an air distribution system. Short end runs with *flexible duct* ("flex") for the supply air, for example, often allow minor placement adjustments of diffusers in ceiling grids after a building is occupied. Also, supply air grilles can be specified with movable vanes, which are useful for postoccupancy adjustments.

3.2.4. Ventilation, Air Change, and Contaminant Removal Effectivenesses

Ventilation effectiveness is the ability of an air distribution system to dilute and remove an in-space-produced contaminant and is thus highly relevant to ETS design. *Air change effectiveness* is the ability of the system to deliver ventilation air to occupants and is also relevant. The *contaminant removal effectiveness* is the ability of an exhaust system to extract a contaminant. Values of 1.0 for these effectiveness measures indicate performance equal to perfect mixing (ASHRAE 2001, ch. 26). A value less than 1.0 indicates less-than-perfect mixing performance and can occur for entrainment flow-type systems, particularly when in heating mode with ceiling-mounted outlets and inlets. But, as stated previously, well-designed systems can approach effectivenesses of 1.0, and values of 0.85 to 0.90 are common for good ceiling-based entrainment flow systems.

Effectivenesses greater than 1.0 is possible, usually via supply air outlets near the occupants for the air change effectiveness or by placing exhaust air inlets near the contaminant sources for the ventilation effectiveness. Displacement flow, which can sweep pollutants away from occupants and deliver ventilation air with little prior mixing with "used air," can achieve values greater than 1.0.

A significant problem with these effectiveness values is that they are very difficult to measure. They require that the space of interest already be built and the HVAC equipment and furnishings be installed. The needed tracer-gas test equipment and personnel are fairly expensive, and the tests can be time consuming and intrusive. Many variables are hard to control in these field experiments, such as the thermal loads and occupancy, so obtaining repeatable results is very difficult. Measuring ventilation and air-change effectivenesses in lab mock-ups is possible (e.g., via ANSI/ ASHRAE 1997 [RA 02]; Standard 129), as is making first-order predictions of system performance with theoretical or numerical models. Software models such as *computational fluid dynamics* (*CFD*) programs are promising but are currently too resource-intensive for everyday HVAC design purposes. But an official interpretation of Standard 62 (62-2001- 06) advises that if the *air diffusion performance index* (*ADPI*) of an air diffusion system is greater than 75, then the ventilation effectiveness can be assumed to be 100% (=1.0). The ADPI, and how to evaluate it, is presented in Rock and Zhu (2002) and the *ASHRAE Handbook* (ASHRAE 2001, ch. 32). The recently adopted Addendum *n* to Standard 62-2001 includes in its Table 6.2 "zone air distribution effectiveness" values for various room airflows, and this also appears in the recently published revision, Standard 62.1-2004.

3.2.5. Air Cleaning

One previously mentioned way to create ventilation air is to provide air treatment—not only of recirculated air but also of less-than-suitable outside air. In-space air treatment is possible, but such units are often undersized or otherwise poorly selected or noisy and may be subject to occupant interference. The most capable air-treatment devices are normally incorporated into or added onto the central HVAC systems, either during the original design or as retrofits. "Air treatment," for ETS and similar applications, implies removal of particulate and gaseous contaminants from the air and thus "cleaning." In applications where outside air is not available, such as in spacecraft, submarines, and sealed vehicles typical for battlefield or emergency use, air treatment also implies doing this function and scrubbing of CO_2 as well. Air cleaning for ETS applications will be discussed further in Chapters 5 and 6, but much more information on air cleaning can be found in the *ASHRAE Handbook* (ASHRAE 2004, ch. 24) and manufacturers' literature.

3.2.6. ASHRAE Standard 62

ASHRAE Standard 62 was originally developed, and has been revised through the years, to provide guidance to HVAC designers on ventilation rates. In the late 1970s and early 1980s, energy consumption in buildings was internationally a key area of concern, and ventilation was identified as an area of potential savings. Since the early 1980s, HVAC designers have placed more emphasis on achieving the appropriate balance between energy conservation and indoor-air quality. ASHRAE Standards 62 and 90 address these two sometimes competing issues.

ASHRAE Standard 62 has, and still does include, a table of required ventilation air-flow rates. This table's values assume perfect mixing, and adjustments are required for expected air-flow patterns. In the early editions, the minimum ventilation rate for each type of occupancy listed in the table assumed a moderate amount of smoking, presumably a lower percentage for general office spaces and up to 100% for smoking lounges. The standard typically required 10 to 20 CFM (5 to 10 LPS) per person of ventilation air to achieve acceptable IAQ in general occupancies.

However, due to energy consumption concerns, the standard's Table 2 values were reduced in 1981 to as low as 5 CFM (2.5 LPS) per person in nonsmoking spaces. Complaints arose about indoor-air quality and "sick buildings," and part of the blame at the time was directed at these reduced ventilation rates. People's knowledge about, and expectations for, air quality have increased too, and later revisions of Standard 62 restored the required ventilation air rates to the previous or even higher levels.

As health and comfort concerns about ETS have increased, ASHRAE Standard 62 required even further revisions. As of this writing, the current version of the standard for mainly nonresidential applications is *ANSI/ ASHRAE Standard 62.1-2004*, and it includes many previously approved addenda.

While it is hard to predict the future, one thing is now certain about Standard 62—it has been split into two parts. One portion, renamed 62.1, addresses ventilation of commercial, institutional, and high-rise residential buildings, and recently published 62.2 concerns ventilation in low-rise residential buildings. The residential 62.2 doesn't specifically address ETS at all, but instead discusses occupant-controlled "high-polluting events" in general (ASHRAE 2004b).

Recent versions of Standard 62, now 62.1, have included two methods for determining ventilation air-flow rates. The *ventilation rate procedure* prescribes specific ventilation air-flow rates, and Table 2, now 6.1 via the recently published addendum *n*, of the standard is the most widely referenced part of that procedure. The rates given in Table 2, or 6.1, are the amounts of ventilation air that are to be *delivered* to the occupants and not just the outside air-flow rate entering an air handler. Because Table 6.1 assumes perfect mixing, with near-zero leakage from the supply ducts, these values can in theory be similar. But in practice, especially with VAV systems relying on outside air for their ventilation air, the flow rate through the *OA* louvers may need to be significantly higher than the base values given in Standard 62's Table 6.1, especially when the air-change effectiveness of the air distribution system is expected to be well below 1.0. Addendum *n* provides adjustment methods.

The other method in Standard 62 for setting ventilation air-flow rates is performance based, rather than prescriptive, and is commonly known as the *IAQ procedure*. This method requires identifying the contaminants of concern, examining the quality of the available ventilation air, determining the needed ventilation rates to keep the contaminants at the levels needed for acceptable indoor-air quality, and testing to make sure that the objectives have been met.

In practice, the ventilation rate procedure has been the overwhelming method used by HVAC designers due to the difficulties of complying with the requirements of the IAQ procedure. However, as the standard has continued to be revised, more information on how to apply the IAQ procedure is being added, and additional IAQ procedure-like requirements have been included in the prescriptive method. For example, designers are now required to at least observe and then document the quality of the outside air when using the prescriptive method (62-2001; addendum *r*). In addition, another official Standard 62 interpretation (62-2001-10) and addendum *o* indicate that more ventilation air be provided when ETS is present; just using the base ventilation rate procedure alone is not appropriate. One more interpretation (62-2001-17) says that filtered recirculated air can be used to provide ventilation air and lower the admission rate of outside air, but the IAQ procedure must be used. Completed and pending changes to Standard 62 have substantially removed consideration of ETS, however, from 62.1. In the next chapter, various methods for estimating the needed ETS ventilation rates will be presented.

3.3. ADAPTED VERSUS UNADAPTED

In regard to the perception of both thermal comfort and acceptable indoor-air quality, people's opinions vary depending on how long they have been in a particular environment. Someone who has resided in a space for a while may become more accustomed to his or her surroundings and can then have a higher threshold to continued thermal discomfort or reduced objections to some odors, for example. These *occupants*, due to their

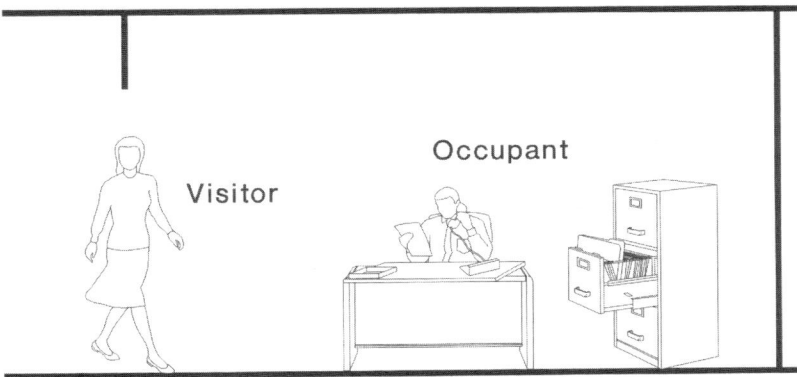

Figure 3.8. An "occupant" is expected to be within a space for a long time and is normally considered "adapted" to odors. Visitors, having not been exposed recently, tend to be more sensitive to odors and irritants and are considered "unadapted."

reduced sensitivity, are called *adapted* building users, as shown in Figure 3.8. Their condition, in regard to some odors, is called "olfactory fatigue" (Knutson 2003). However, individuals who have not been in the space, or have been away from it for some time, are more sensitive upon entry. These building users are known as *visitors* or *unadapted* people.

A particular visitor who initially senses discomfort, odors, or irritations, for example, may later feel the environment is acceptable even though nothing may have changed with the HVAC system or its settings. This adaptation is caused by many physiological effects such as reduced metabolic rate—for example, going from walking to seated; changes in clothing level; and the olfactory "numbness" that develops to unpleasant odors (ANSI/ASHRAE 1992; Fanger 1972). How long does this adaptation require? The studies vary, but 15 minutes seem to be a reasonably conservative value for many odors. However, irritations may get worse, rather than better, as a person stays in a space. Irritations and adaptation are matters for continuing research efforts.

Given this difference in adapted versus unadapted perceptions, for which should we design? It depends on each situation. For example, if designing an HVAC system for an office building that has workers present all day long, but few customers stopping in, designing only for the occupants may make more sense. The same "adapted" conclusion might be drawn for long-term occupants of entertainment facilities, such as bars, nightclubs, and casinos, for example. For a space that has many short-term visitors, such as the lobby of a hotel, designing for the unadapted people is probably more appropriate. In mixed-use spaces, where some ETS may be present, judgment is needed. For example, when a restaurant's waiting area is next to both smoking and nonsmoking sections, designing the waiting area for unadapted guests seems best—this way fewer nonsmokers may change their minds about frequenting the establishment. In longer-term occupancies, such as jails or casinos, designing for adapted occupants may be appropriate. Examples of adapted versus unadapted ventilation rates appear in the following chapters.

3.3.1. Smokers and Nonsmokers

Not surprisingly, on average, smokers are much more tolerant of ETS odors and irritants than nonsmokers. Thus, for odor and irritation control only ventilation air-flow rates for spaces that are planned to house only smokers might be much lower than those for mixed occupancy and may still achieve 80% or so satisfaction. Three or more times the ventilation air may be needed to satisfy nonsmokers (Cain et al. 1983). The third method presented in the next chapter for estimating ventilation rates takes both the adapted/unadapted and smokers/nonsmokers occupancy factors into account.

4

VENTILATION RATES FOR ETS

The air-flow information provided in this book may at times look somewhat like the prescriptive ventilation rate procedure of ANSI ASHRAE Standard 62.1, but at other times it may sound more like the performance IAQ procedure. As such, the guidance that follows does in places lack the preciseness desired from a prescriptive method, but it is hoped that through the information presented, and using critical thinking based on education and experience, you will be able to design HVAC systems that appropriately handle ETS's odors and irritations while understanding that health concerns may persist.

In this chapter, several different approaches to determining ventilation rates for spaces with environmental tobacco smoking are presented. Over the years there have been a number of studies of the ventilation rates for acceptable IAQ by various research groups and practitioners. Some have concluded that for secondhand smoke the required outside air-flow rates are extremely high, and thus not practical, while others have determined that even the base, non-ETS rates in Standard 62.1 are conservative even when applied directly to ETS-allowed spaces. The three rate estimation methods presented in this chapter show these three common, yet significantly different, design philosophies. They are: theoretical rate with perfect mixing, ventilation for ETS in aircraft, and the ETS dilution method for buildings.

4.1. THEORETICAL RATE WITH PERFECT MIXING

Ventilation rates for ETS have been derived through theoretical calculations based on the number of occupants, the percentage that are smokers, the average rate of smoking per user, and the rate of contaminant production per unit tobacco product in use. Most of these calculations assumed that only cigarettes were being smoked and that the room air was perfectly mixed. The following derivation and a sample calculation use such an approach too. For the example provided, only a low concentration of ETS is allowed to demonstrate the method's predicted high ventilation rate.

4.1.1. Conservation Equations

Figure 4.1 shows graphically the conservation of mass applied to a simple room; the room shown has one air outlet and one inlet; no recirculation is assumed, so it has a "once-through" ventilation system. If the mass flow rate of interest is that of a particular contaminant—for example, ETS—then the mass flow rates become

$$\dot{m} = \rho \cdot \dot{V} = C \cdot \dot{V} \qquad (4.1)$$

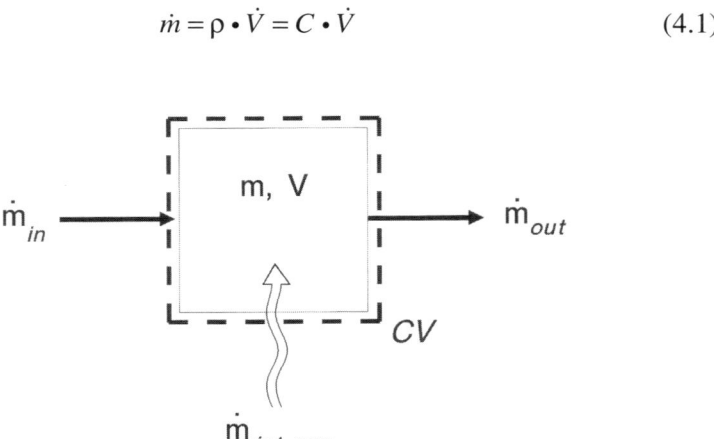

Figure 4.1. The conservation of mass for a control volume. The internally generated mass flow is shown crossing the control volume's boundary for simplicity.

In equation 4.1, the mass flow rate ($\dot{m}$) is in lbm/min (kg/s), the density (ρ) is in lbm/ft^3 (kg/m^3), and the volumetric flow rate ($\dot{V}$) is in ft^3/min (m^3/s). V in Figure 4.1 is the volume of the room in ft^3 (m^3), and the variable with the "int gen" subscript is for the internally generated contaminant of concern—ETS in this case.

Because relative quantities of airborne contaminants are normally described with their concentrations (C), the density can be replaced with C in equation 4.1. If only one chemical species is being tracked, it is often called a tracer or marker. Concentrations of materials can be expressed in many ways: parts per million parts (ppm), % by mass or volume, or, as in equation 4.1, their mass per unit volume. When concentrations are tracked, as shown in Figure 4.2, the mass balance of that substance is often called the conservation of species, and for this control volume (cv) is

$$C_{in} \bullet \dot{V}_{in} - C_{out} \bullet \dot{V}_{out} + C_{int\ gen} \bullet \dot{V}_{int\ gen} = V \bullet \left.\frac{dC}{dt}\right|_{cv} \qquad (4.2)$$

If we assume that the volumetric flow rate of the internally generated contaminant is very small, as compared with the air-flow rate, and this is a good assumption for typical HVAC design and ETS, then the air-flow rate into the control volume is equal to the air-flow rate out, so

$$\dot{V}_{in} = \dot{V}_{out} = \dot{V} \qquad (4.3)$$

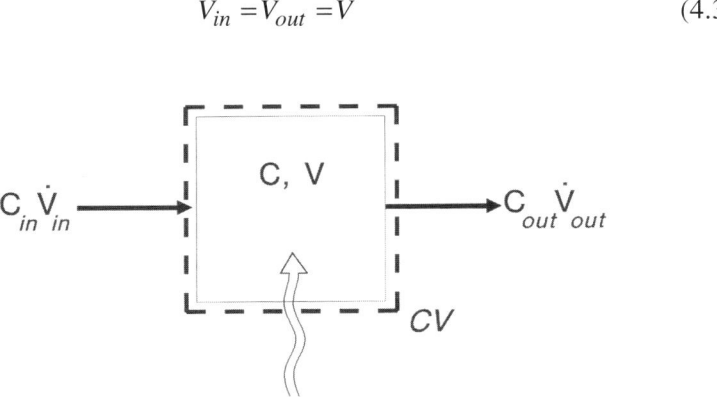

Figure 4.2. The masses and mass flow rates can be replaced by the density or concentration and the volumes or volumetric flow rates of the substance of interest. For ETS, the internal generation is still normally reported as a mass flow rate, however.

For a once-through, 100% OA system, the inlet concentration of the contaminant is the concentration in the outside air (ambient level), so

$$C_{in} = C_{oa} \approx constant \tag{4.4}$$

If the internal generation of the contaminant is zero at and before time $= 0$, and a long time period has elapsed before $t = 0$ so that the indoor concentration is the same as that outdoors,

$$C(t = 0) = C_{out}(t = 0) = C_{oa} \tag{4.5}$$

Immediately after time zero, the internal generation of the contaminant begins: the smokers start smoking, and the concentration (C) in the room begins to rise, as does C_{out}. In a real room, the surfaces would begin a net ab- and adsorption of the contaminant, so they would slightly delay the rise in concentration. But as time goes to infinity $(t \rightarrow \infty)$, it is assumed that the rates that materials ad- and absorb become equal to the rates at which they readmit the contaminant, so the net effect is that eventually the surfaces no longer remove the contaminant from the room air. These rates vary for the different components in ETS and other gaseous mixtures. The concentration in the perfectly mixed room reaches steady-state as time goes to infinity, so the transient term in equation 4.2 goes to zero. And because of the perfect mixing assumption, the concentration of the exhaust air is the same as the room, or $C = C_{out}$, so

$$C_{oa} \bullet \dot{V} - C(t \rightarrow \infty) \bullet \dot{V} + \dot{m}_{int\ gen} = 0 \tag{4.6}$$

Solving for the steady-state room air concentration $(C(t \rightarrow \infty))$ gives

$$C(t \rightarrow \infty) = C_{oa} + \frac{\dot{m}_{int\ gen}}{\dot{V}} \tag{4.7}$$

or

$$C(t \rightarrow \infty) = C_{oa} + C_{int\ gen} \bullet \frac{\dot{V}_{int\ gen}}{\dot{V}} \tag{4.8}$$

These two equations can be rearranged as

$$\dot{V} = \frac{\dot{m}_{int\ gen}}{C(t \to \infty) - C_{oa}} = \frac{C_{int\ gen} \cdot \dot{V}_{int\ gen}}{C(t \to \infty) - C_{oa}} \qquad (4.9)$$

Equation 4.9 can be used to determine the outside air-flow rate needed to maintain the contaminant concentration at or below a desired level ($C(t \to \infty)$) for the given assumptions. In determining the flow rate, consideration is needed for control of other contaminants too—for example, building materials off-gassing, occupant odors, and so on—so the overall rate may vary.

Most recent ventilation studies of ETS concentrations and/or health effects use this first method (Walker 1997; Nazaroff and Singer 2002), and it will likely continue to be the method of choice for future related research. The studies typically propose that very high air-flow rates are needed for acceptability, or that no reasonable ventilation rates can reduce ETS, or any particular component of interest, to "safe" levels (Glantz and Schick 2004).

4.1.1.1. Sample Flow-Rate Calculation

In this section, the outside air rate needed to ventilate a room with one smoker present is found using the preceding method. The result will be a CFM/person rate, so it can be scaled up for multiple smokers. The assumptions in Section 4.1.1 are made for this room, so, for example, ceiling-based air diffusion with an ADPI of 85 was selected and thus near-perfect mixing is assumed.

As the building in which this smoking-allowed room is well away from other sources of ETS, the outdoor concentration is likely very low and thus assumed to be zero, so the rate equation, equation 4.9, becomes

$$\dot{V} = \frac{\dot{m}_{int\ gen}}{C(t \to \infty)} = \frac{C_{int\ gen} \cdot \dot{V}_{int\ gen}}{C(t \to \infty)} \qquad (4.10)$$

The first of two challenges in solving this simple equation is determining the rate of emission of ETS products. Due to the variation of products, how users smoke, and the randomness of occupancy and percentage of smokers, it is probably not possible to determine a highly accurate rate for all situations in the real world. But many attempts have been made, and success in determining rates in controlled conditions has been achieved. Repace et al. (1998),[1] using data from Leaderer and Hammond (1991) for ten different major brands of cigarettes and one cigar, predict an RSP rate of 2.27 mg/min per smoker. For comparison, Table 3 of Chapter 44 of the 1999 volume of the *ASHRAE Handbook* (ASHRAE 1999) gives a much lower RSP value of 13.674 mg/cig, or about 1.82 mg/min per smoker. VOCs and other gases produced are not included in this RSP value. The average time to smoke a cigarette was 7.5 minutes.

The second challenge is setting the concentration limit (*C*). If U.S. regulatory agencies had set acceptable exposure limits, the most appropriate value would be used here. Lacking such, in Table C-4 of ANSI/ASHRAE Standard 62-1999, whose information was extracted from a World Health Organization report (WHO 1986), a "concentration of limited or no concern" is stated as <0.1 mg/m^3 of RSPs and a "concentration of concern" as >0.15. A remark in the table states that, as of that time, the Japanese standard was 0.15 mg/m^3. Using the lower 0.1 as only an example and not an actual limit, and the generation rate of RSPs at 2.27 mg/min per smoker, equation 4.10 gives

$$\dot{V} = \frac{2.27\,mg/min}{0.1\,mg/m^3} = 22.7\,m^3/min = 378\,l/s \approx 801\,ft^3/min \qquad (4.11)$$

As this example shows, a "chain smoker" (continuous smoking and therefore not realistic for long time periods) requires a high rate of ventilation air to maintain a low concentration of RSPs in a perfectly mixed room. This example, when instead evaluated per cigarette (7.5 min × 800 CFM = 6,000 ft^3), gives somewhat comparable results to those described by Cain et al. (1983), who showed that about 4,000 ft^3 per cigarette is needed for 80% satisfaction of odor control. This widely referenced 1983 experimental study used a floor-to-ceiling displacement-like room geometry but with high recirculation rates that provided mixing.

Note that the concentration values 0.1 and 0.15 in Table C-4 of Standard 62-1999 have been deleted from later versions of the standard. The value 0.1 used previously was an example and should not be construed to

---

1. It is not apparent whether or not a monograph such as this has been subjected to a peer-review process.

be a recommended value. An ETS/health research and no-smoking advocacy group has recently suggested that about 0.6 µg/m³ of ETS RSPs *may* be an appropriate value for odor control only, but that unfeasibly high rates of outside air admission are required to achieve this concentration [Glantz and Schick 2004].

4.2. VENTILATION FOR ETS IN AIRCRAFT: HISTORICAL DEVELOPMENT OF METHODS

In the 1991 *ASHRAE Handbook* (ASHRAE 1991, ch. 9.3), a method was presented for determining ventilation rates in aircraft with ETS. Since that time smoking on U.S. domestic commercial flights has been banned. However, a few airlines outside of the United States, primarily in Asia and Africa, still allow smoking onboard their aircraft.

In modern jet aircraft, outside air is normally obtained via air bled off a compressor stage of one or more engines. Until recent decades, the ventilation systems were once-through, but now 50% recirculation is becoming typical. The recirculated air is usually filtered using high-efficiency media. Two situations of having air recirculated or not between the smoking and nonsmoking sections of aircraft were presented in the ASHRAE *Handbook*. Normally, recirculation of ETS-laden air to nonsmoking portions would not be allowed. An exception would be the presence of extremely well filtered recirculated air, as will be discussed later in this chapter and in more detail in the next chapter.

Jet aircraft tend to have ceiling-to-floor or floor-to-ceiling air diffusion, so the rates that follow should be lower than that presented in Section 4.1.1 for perfect mixing, because a degree of displacement flow occurs. But with increasing use of recirculation, jet aircraft will behave more closely to the perfect mixing case. As the example will show, rather low ventilation air-flow rates result from this method. Where the example in the previous section was intended to show the potentially very high rates needed, this section's example shows nearly the other extreme—a very low ventilation rate needed for adapted occupants' satisfaction. A lone reference is mentioned (Thayer 1982) in Chapter 9 of the 1991 *ASHRAE Handbook* for this design method.

While this *Environmental Tobacco Smoke Design Guide* was not envisioned for application to aircraft, the method that appeared in the *Applications* volume, and repeated here, is useful for comparison and the historical development of some of the current design philosophies for ventilating spaces. As the rates in smoking-allowed rows of aircraft assume that only a portion of the smokers are smoking at any one time, care is needed in comparing this section's rates to those for buildings. For example, smoking break-rooms in buildings have similar high densities of adapted occupants, similar to that in the smoking sections of aircraft. However, the number of active smokers will probably be higher in the break-rooms than in the aircraft, as the time in residence is shorter in the break-rooms. The rates for buildings' smoking lounges are discussed further later and should be higher than that recommended by the previous ASHRAE *Handbook*'s "aircraft method," which follows.

4.2.1. Procedure

The method presented in the 1991 *Applications* volume applies to two cases: aircraft with separate smoking and nonsmoking sections, and aircraft with mixed seating throughout. Both are for occupants' perceptions of comfort only. Via experiments mentioned in the volume, values for two indices were found and can be used to estimate the needed ventilation flow rates. These two indices are the dilution index (DI) and the irritation index (II). The dilution index, with units of l/mg, is the volume of air needed to dilute ETS to a certain level of acceptance by occupants. In the underlying experiments, presumably done by a major aircraft manufacturer, values up to about 100 l/mg were studied. The irritation index is a subjective rating between 0 and 5 given by test occupants, and its values are:

0 = imperceptible
1 = threshold
2 = acceptable
3 = annoying
4 = objectionable
5 = intolerable

Figure 4.3 shows the observed relationship between DI and II for both smokers and nonsmokers. As expected, the figure shows that smokers are more tolerant of ETS odors and irritants than nonsmokers.

Of more direct use is Figure 4.4, also from Chapter 9 of the 1991 *ASHRAE Handbook*, which shows occupant satisfaction, expressed as a percentage, versus the dilution index. Separate curves are presented for

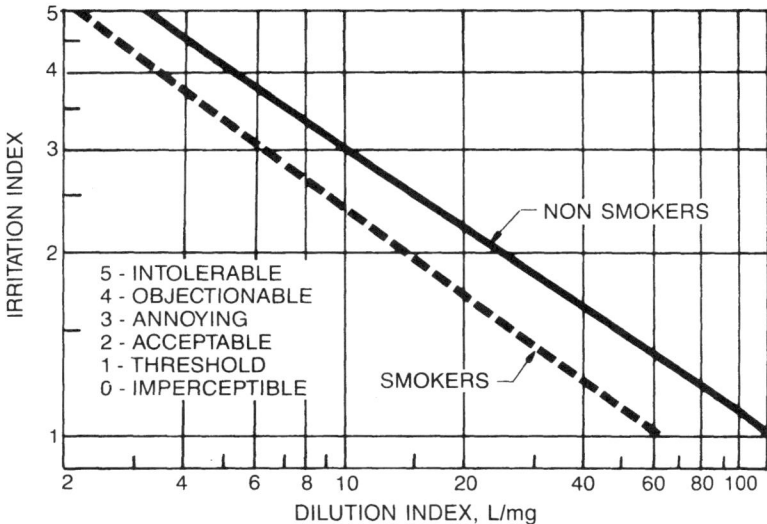

Figure 4.3. How the observed irritation varies with volume of dilution air for ETS in aircraft. As expected, smokers are less sensitive than nonsmokers (ASHRAE 1991, ch. 9.2).

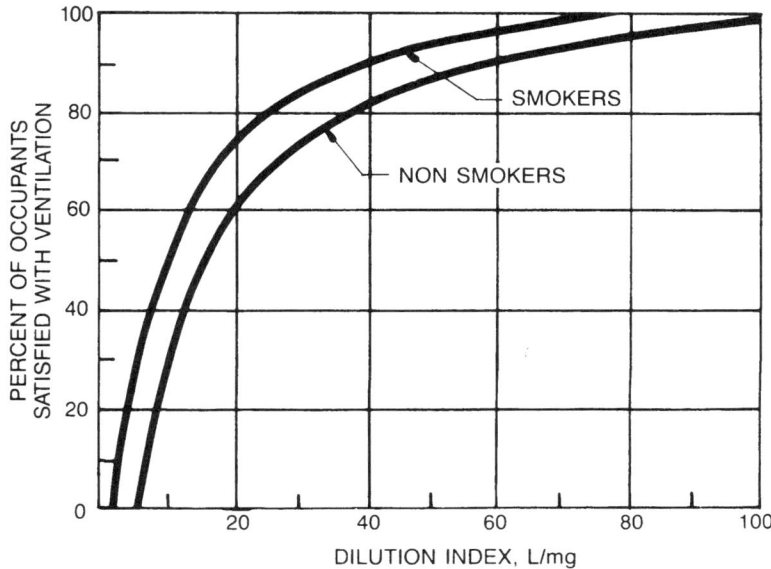

Figure 4.4. Percent of occupants satisfied as compared with the dilution index (ASHRAE 1991, ch. 9.2).

smokers and nonsmokers, and, for example, 80% of smokers find ~24 l/mg dilution acceptable, while nonsmokers require ~37 l/mg. Knutson (2003) advises that "odor control" techniques are intended for the "average" occupant and that reasonable control measures may still yield complaints from those with more acute noses; thus, even smokers, 20% in this case, may complain.

Note that this aircraft method does *not* separate the ventilation air-flow rate needed to dilute and remove ETS contaminants from that needed to handle those from other sources, such as the occupants, luggage, or the aircraft itself. Instead, the method combines the ventilation functions into one base air-flow rate. The method states that a minimum of 15 CFM (7.5 LPS) per person should be provided.

4.2.2. Sample Calculations

For a "smoking zone" some critical pieces of information needed are: the rate that ETS contaminants are produced per smoker, and the likely number of people actually smoking. The example given in the *ASHRAE Handbook* assumes 17 mg/min per smoker, presumably RSPs. As shown previously, a much lower emission rate of 2.27 mg/min has been reported by other researchers. However, using the method's higher value and Figure 4.4 with 80% smoker's acceptance,

$$\frac{24\,\dfrac{l}{mg} \cdot 17\,\dfrac{mg}{min}}{60\,\dfrac{s}{min}} = 6.8\ LPS = 14.4\ CFM \tag{4.12}$$

This 14.4 CFM is per smoker, and not all those in the smoking section of the airplane will likely be smoking at the same time. If 15 CFM per person is the minimum ventilation rate needed for the aircraft anyway, the results indicate that slightly more than 80% of the smokers will be satisfied with that base ventilation rate.

For a "nonsmoking zone" an implicit assumption is that no significant ETS will migrate directly to it from the smoking zone, and the base 15 CFM per person will control odors and irritants from all sources. However, the method does allow for recirculation of air from the smoking zone to the nonsmoking zone if filtered to remove all the ETS or if the recirculated air is sufficiently diluted. A satisfaction rate of 90% for nonsmokers is suggested in the 1991 *ASHRAE Handbook*. From Figure 4.4, for 90% of nonsmokers satisfied, about 60 l/mg of dilution air is required. The method gives the following equations for finding the amount of outside air as compared with recirculated air in the nonsmoking zone:

$$k \equiv \frac{DI_{nonsmokers}}{DI_{smokers}} - 1 \qquad (4.13)$$

and

$$\dot{V}_{oa} = k \bullet \dot{V}_{ca} \qquad (4.14)$$

Continuing the example, with $DI_{nonsmokers} = 60$ l/mg and $DI_{smokers} = 24$ l/mg, the intermediate variable k is 1.5. So, from equation 4.14, the volumetric flow rate of the outside air to the nonsmoking section should be 1.5 times that of the air recirculated from the smoking section of the airplane. Providing separate ventilation systems, which may not be possible due to size and weight concerns in aircraft, may be more effective, because the required air-flow rate to the nonsmoking section could be substantially lower. The 1995 version of the *ASHRAE Handbook* removed the part of the method that allowed for recirculating air from the smoking section to the nonsmoking (ASHRAE 1995, ch. 9).

For aircraft that have mixed seating, typical of smaller and often private aircraft, the method recommends design using the flow rate needed to produce 90% satisfaction of the nonsmokers. For 60 l/mg and a 17 mg/min ETS rate, the method gives

$$\frac{60\,\frac{l}{mg} \bullet 17\,\frac{mg}{min}}{60\,\frac{s}{min}} = 17\ LPS = 36\ CFM\ per\ smoker \qquad (4.15)$$

Not knowing where or how many smokers there will be, a somewhat conservative conclusion is that 36 CFM should be supplied for every occupant. The method recommends that aircraft with mixed seating should have well-mixed air diffusion. The 1995 version of the method deleted these recommendations for mixed seating (ASHRAE 1995, ch. 9).

By the 1999 version of the *Applications* volume, the method was gone entirely (ASHRAE 1999, ch. 9).

4.2.3. Current Aircraft Recommendations

The current version of the *Applications* volume has a brief section on ETS in aircraft and suggests a Standard 62 IAQ procedure-like approach to selecting ventilation rates. It also states that via a Department of Transportation study of 92 aircraft with smoking sections, the average RSP measurements were 0.040 mg/m^3 in the nonsmoking sections and 0.175 mg/m^3 in the smoking (ASHRAE 2003b, ch. 10.8). A separate ASHRAE standard, 161P, for air quality within commercial aircraft has been in development for many years by its Standard Project Committee (SPC).

4.2.4. Other Nonbuilding Applications

Besides aircraft, there are other applications that may or may not fall under the intended guise of ASHRAE Standard 62. For example, ocean-going vessels fall under the purview of the U.S. Coast Guard and its regulations; even floating riverboat gambling casinos may be under the Coast Guard's auspices, because these ships or their ancestors once made runs through the open ocean to reach neighboring ports. Requirements for land-going vehicles, such as cars, trucks, and buses, are covered by various federal, state, and private organizations. Military and some other government-operated fixed and mobile structures, and sensitive applications in the nuclear, petrochemical, and medical industries, for example, may have significantly different requirements. But where no considerations previously exist, or requirements are being revised, more references are being made to Standard 62, for example, by the controlling authorities for these nonbuilding applications.

4.3. ETS DILUTION METHOD
FOR BUILDINGS

This section presents another method for determining ventilation rates and often leads to results between those from the previous two methods. The ETS Dilution Method (EDM), a methodology deleted from an informative appendix in a previous version of ASHRAE Standard 62.1, can be used to determine the ventilation rates for odor and irritation control when ETS is present. It is an applied version of the first method presented in this chapter and has been enhanced by others with the factors and input values needed for HVAC design purposes. The EDM is similar to some material in Nelson et al. (1998) and Bohanon et al. (1998). While some argue that the EDM results in insufficient rates, others are equally convinced that the EDM rates are far in excess of what are needed for odor and irritation control. Both may be correct from their points of view; you, as the designer, must select the most appropriate input values for your project—the EDM's results depend on them. The ETS Dilution Method is intended to predict the needed ventilation rates for comfort in real spaces and occupants' behavior, but it is much more conservative than the ventilation rates needed for just 75% or so satisfaction of adapted, smoking occupants.

4.3.1. Additivity

A base assumption for the EDM is that Standard 62.1's Table 6.1, formerly Table 2 of Standard 62, adjusted prescriptive ventilation rates needed to ventilate spaces for their normal use ($\dot{V}_{vrp}$), and an additional and often substantial amount of ventilation air is needed to handle the dilution and removal of ETS ($\dot{V}_{ets}$), so

$$\dot{V}_{tot} = \dot{V}_{vrp} + \dot{V}_{ets} \qquad (4.16)$$

A significant concern about this "additivity" approach (Bluyssen and Cornelissen 1999) is that no credit is given for the base ventilation air's ability to also help remove the ETS, and thus spaces may be over ventilated. But the resulting, higher-than-minimal values should decrease the percentage of occupants that perceive unacceptable levels of ETS odors and irritants. Internally generated non-ETS contaminants should be diluted and removed much better as well but likely at a cost of greatly increased energy consumption, or at least more frequent replacements of special and typically expensive filters.

A more aggressive application of this method would be to reject additivity as some empirical studies' findings suggest (Bohanon et al. 2003). The designer would find both $\dot{V}_{vrp}$ and $\dot{V}_{ets}$ but use only the higher of the two. The examples that follow, however, assume that the more conservative additivity approach is retained.

4.3.2. Extra Ventilation for ETS

The basic formula for determining the required extra ventilation air, $\dot{V}_{ets}$, follows in equations 4.17a for I-P units and 4.17b for SI. In the equations, $\dot{D}_{cig}$ is the design smoking density that is the estimated rate of cigarettes (cig) being smoked each hour per unit occupiable floor area for the space in question (cig/h·ft^2 [cig/h·m^2]). Suggested values for $\dot{D}_{cig}$ will be reported later. The specified volume of ventilation air needed to dilute each cigarette's products is V_{cig} and has units of ft^3/cig (m^3/cig). An equation and a table of values for V_{cig} will also be given later. *A* in equations 4.17a and b, and Table 2 of Standard 62, is the *net occupiable* floor area of the space in ft^2 (m^2), and not the total floor area. The *contaminant removal effectiveness*, E_{cr}, has a value of 1.0 for perfect mixing, less than 1.0 for entrainment flow with poor mixing, and can be greater than 1.0 for displacement flow. As previously discussed, if the ADPI will be 75% or above, E_{cr} is often assumed to be 1.0.

$$\dot{V}_{ets} = \frac{\dot{D}_{cig} \bullet V_{cig} \bullet A}{60\,\dfrac{min}{h} \bullet E_{cr}} \qquad \text{(I-P units)} \qquad (4.17a)$$

$$\dot{V}_{ets} = \frac{\dot{D}_{cig} \bullet V_{cig} \bullet A}{3.6\,\dfrac{s \bullet m^3}{l \bullet h} \bullet E_{cr}} \qquad \text{(SI units)} \qquad (4.17b)$$

4.3.3. Air Volume per Cigarette (V_{cig})

In this section, the volumes of ventilation air needed to dilute the ETS to acceptable odor and irritation levels are presented. These values are based on the research of Leaderer et al. (1983) and others. Questions exist on the current relevance of these data given changes in the prevalence of smoking and perhaps in people's expectations, but no other values are available at this time. As the presented dilution volumes are not currently codified, it is up to you as the HVAC designer to identify and select appropriate values, and they may differ significantly from those presented here.

The ventilation air represented by V_{cig} is normally provided through outside air, but highly cleaned recirculated air may be used for odor control to partially, or even fully, meet this need, as will be discussed further. If cigars are being smoked, the EDM describes that each should be considered as equal to four cigarettes, so substantially more ventilation air will be needed. Tables 4.1a for I-P units and 4.1b for SI units present the needed volumes for smokers and nonsmokers, as well as unadapted (visitors) and adapted (occupants) people.

The data for V_{cig} are based on chamber studies that included certain percentages of volunteer smokers (*sm*) and nonsmokers (*ns*). It is possible that the percent nonsmokers will increase in a particular space over time as society evolves, and thus more ventilation air will be needed to satisfy a certain fixed percentage of the total occupants. The data also assume perfect mixing, and only healthy, adult volunteers were used in the testing. Specifying higher ventilation air volumes than those presented in Tables 4.1a and b may be prudent (Glantz and Schick 2004), or at least selecting system components with enough spare capacity so that the rates could be increased somewhat in the future may be wise. But because the underlying experiments for the air volumes were done in a plain metal test chamber, distractions in real rooms may cause occupants to be less sensitive, and ab- and adsorption may decrease concentrations; lower air volumes per cigarette than those presented may also prove acceptable.

4.3.3.1. V_{cig} for Mixed Occupancies

The data in Tables 4.1a and b are for four specific cases, one being "adapted" and "smokers." When there is a combination of occupancies, such as a mixture of smokers and nonsmokers in the same room, the volume of ventilation air needs to be adjusted. The needed ratio of smokers to total occupancy (X_{sm}) is

$$X_{sm} = \frac{P_{sm}}{P_{tot}} = \frac{P_{sm}}{P_{sm} + P_{ns}} \quad (4.18)$$

where P_{sm} is the expected number of smokers, P_{ns} is the nonsmokers, and P_{tot} is the total estimated occupancy of the space. Note that later per person (p) values are different from these absolute numbers of occupants (P), or the occupancy densities—that is, the number of people per 1,000 ft^2 (100 m^2)—similar to those presented in Table 6.1, formerly Table 2, of Standard 62. The ratio of smokers, X_{sm}, is technically a dimensionless value, but it is convenient to give it units of p_{sm}/p.

When multiplied by 100%, the smoker-to-total occupancy ratio is the percent smokers. In mixed, general U.S. occupancies, 20% to 30% smokers have been common. Some guidance on this ratio for various spaces can be found in Table 4.2 and elsewhere (e.g., Glantz and Schick 2004). These ratios may be significantly higher in other countries, especially if largely male occupancies are expected, but may over time fall here and abroad if antismoking campaigns continue to be effective.

With the ratio of smokers determined, the adjusted volume of air can be found from

$$V_{cig} = X_{sm} \cdot V_{cig,sm} + (1 - X_{sm}) \cdot V_{cig,ns} \qquad (4.19)$$

where values for $V_{cig,sm}$ and $V_{cig,ns}$ are obtained from Table 4.1a or b.

Table 4.1a. Ventilation Air (V_{cig}) Required per Cigarette Smoked (in I-P units of ft^3/cig)

Occupants	Unadapted	Adapted
Nonsmokers ($V_{cig,ns}$)	5,600	3,900
Smokers ($V_{cig,sm}$)	1,400	1,100

Table 4.1b. Ventilation Air (V_{cig}) Required per Cigarette Smoked (in SI units of m^3/cig)

Occupants	Unadapted	Adapted
Nonsmokers ($V_{cig,ns}$)	160	110
Smokers ($V_{cig,sm}$)	40	30

Table 4.2. Typical Values of Fraction of Occupants Who are Smokers (X_{sm}) and the Smoking Rate per Smoker ($\dot{R}_{sm}$)

Occupancy Type	X_{sm} (dimensionless)	$\dot{R}_{sm}$ (cig/p$_{sm}$·h)
Smoking Lounges	1.00	3–6
Bars, Cocktail Lounges, Casinos, Lunchrooms	0.25–0.50	1–2
All others, U.S. average	0.20–0.25	0.6

The EDM does not make adjustments for the expected ratio of men versus women. A higher percentage of men than women tend to smoke in many countries. The difference is fairly minor in the United States, but in some other countries, such as in Asia, it is dramatic. Therefore, it would be prudent for spaces that will have a mixture of smokers and nonsmokers to adjust X_{sm} or $\dot{V}_{ets}$ up or down depending on the expected proportions of men and women present. For example, a smoking-allowed space in an all-female social clubhouse might need less ventilation air due to a lower X_{sm} to achieve 80% acceptance than a similar space in an all-male equivalent with a higher X_{sm}. Research is still needed, but a counterargument might be that women may be more sensitive, on average, to ETS than men, and may therefore want *more* ventilation air.

4.3.4. Smoking Rates per Smoker ($\dot{R}_{sm}$)

The EDM, and similar emission/dilution methods, requires an estimate of the average rate of cigarette use per predicted occupant. For the EDM, this rate, $\dot{R}_{sm}$, is for smokers who may be present. This average rate can be found by dividing the number of cigarettes consumed (*cig*) by the elapsed time (Δt) in which this occurs. Table 4.3 gives some example values, and the last column of Table 4.2 shows suggested rates from Standard 62-

2001's now-removed Appendix I for the EDM. Some have strongly argued that Table 4.2's values, especially the 0.6 rate, are too low (e.g., Glantz and Schick 2004); other think some values are too high. You will need to estimate appropriate values for each of your designs via suggestions in this book, other references, and from your experiences.

4.3.5. Design Smoking Density ($\dot{D}_{cig}$)

The design smoking density is the rate of cigarettes smoked per hour and per unit occupiable floor area (cig/h·ft^2 [cig/h·m^2]). Note that this value is *not* per smoker but is an overall rate in the EDM. $\dot{D}_{cig}$ can be estimated through

$$\dot{D}_{cig} = \frac{X_{sm} \cdot P_{tot} \cdot \dot{R}_{sm}}{A} \qquad (4.20)$$

where:

X_{sm} is the proportion of smokers (from equation 4.18 or Table 4.2, dimensionless or p_{sm}/p)

P_{tot} is the total estimated maximum occupancy in the space (number of people, p)

$\dot{R}_{sm}$ is the smoking rate per smoker (from Table 4.2 or 4.3, cig/p_{sm}·h)

When available, actual information should be used in calculating the design smoking density. If specific occupancy data are not available, the estimated maximum occupancy density values from Table 6.1, formerly Table 2, of the latest version of Standard 62.1 can be used. As post-2001 versions of Table 6.1 will not have data for smoking lounges, the previous versions had 70 people per 1,000 ft^2 or 100 m^2 of occupiable floor area as their estimated maximum occupancy density.

The following is an example calculation for the equations presented in this section. It is intentionally for a nondescript, generic space and shows, among other things, how the units work. Examples for specific types of spaces will be presented in Chapter 6.

Table 4.3. Estimating Average Values of the Smoking Rate per Smoker ($\dot{R}_{sm}$). Twenty Cigarettes per Pack are Assumed.

Cigarettes (*cig*)	Equivalent Packs	(Δt, hours)	$\dot{R}_{sm}$ (cig/ps$_m$·h)
2	0.1	8	0.25
5	0.25	8	0.63
10	0.5	8	1.3
20	1	8	2.5
5	0.25	12	0.42
10	0.5	12	0.83
20	1	12	1.7
40	2	12	3.3
5	0.25	16	0.31
10	0.5	16	0.63
20	1	16	1.3
40	2	16	2.5

EXAMPLE 4.1

Using the EDM with the additivity approach, what is the ventilation air-flow rate ($\dot{V}_{tot}$) needed for a generic 2,000 ft^2 smoking-allowed indoor work area on the West Coast? A ceiling-based air diffusion system with a predicted ADPI of about 80 has already been selected.

Solution:

The first step is to find the base ventilation air-flow rate from Standard 62. Because the ADPI is greater than 75, the space is assumed to be well mixed, and the ventilation air rates from Table 6.1 of Standard 62.1 may be used directly. For this generic example let's say a value of $\dot{V}_{vrp}/p = 15$ CFM/p is found. The architect's good estimate for the occupancy of the space is 10 people per 1,000 ft^2, so the ventilation rate procedure's base flow rate ($\dot{V}_{vrp}$) is

$$\dot{V}_{vrp} = 2,000 \; ft^2 \cdot \frac{10 \; p}{1,000 \; ft^2} \cdot 15 \frac{CFM}{p} = 300 \; CFM \qquad (4.21)$$

This 300 CFM is the ventilation air required to ventilate the space for all the nontobacco sources of contaminants. Now the flow rate needed for the ETS must be found and added to this base value.

The next step toward this goal is to estimate the percent of occupants who are smokers. The space is for construction in the West Coast of the United States; this region has a low percentage of smokers, on average. Also, the space is not a smoking break-room or bar, so a value of 20% smokers is assumed, because Table 4.2 gives a range of 0.2 to 0.25 for X_{sn} for "all other occupancies" in the United States. If the percent of smokers changes over the life of this completed project, the space may prove to be overventilated from an odor and irritation control perspective.

Because this is a work area, the occupants in this generic example will be assumed to be long-term and thus adapted. Table 4.1a then provides the volumes of air needed to dilute the smoke as $V_{cig,ns} \approx 3,900 \; ft^3/cig$ and $V_{cig,sm} \approx 1,100 \; ft^3/cig$. Using equation 4.19, the adjusted volume of dilution air per cigarette is then

$$V_{cig} = 0.2 \cdot 1,100 + (1-0.2) \cdot 3,900 = 3,340 \frac{ft^3}{cig} \qquad (4.22)$$

Tables 4.2 and 4.3 provide guidance on the smoking rate, and a value of $\dot{R}_{sm} = 0.6 \; cig/p_{sm} \cdot h$ is selected from Table 4.2 for this occupancy. As the total occupancy (P_{tot}) is 20 p, the design smoking density is, therefore,

$$\dot{D}_{cig} = \frac{0.2 \frac{p_{sm}}{p} \cdot 20 \; p \cdot 0.6 \frac{cig}{p_{sm} \cdot h}}{2,000 \; ft^2} = 0.0012 \frac{cig}{h \cdot ft^2} \qquad (4.23)$$

As the space is well mixed, the contaminant removal effectiveness, E_{cr}, is assumed to be 1.0. From equation 4.17a, the extra ventilation air needed to dilute the ETS to acceptable odor and irritant levels is then

$$\dot{V}_{ets} = \frac{0.0012 \frac{cig}{h \cdot ft^2} \cdot 3,340 \frac{ft^3}{cig} \cdot 2,000 \, ft^2}{60 \frac{min}{h} \cdot 1.0} \approx 134 \, CFM \qquad (4.24)$$

The total ventilation air-flow rate required for this space, from equation 4.16, is then

$$\dot{V}_{tot} = 300 \, CFM + 134 \, CFM = 434 \, CFM \qquad (4.25)$$

or about 145% of the base requirement for a similar nonsmoking space. As will be seen in later examples, the percent increase over the base VRP rate will be dramatically higher for some spaces—for example, for those with greater occupancy densities and percentages of smokers.

There are significant published research results, typically using test chambers for their experiments and not real spaces, that show the air volume needed per cigarette; the smoking densities may need to be somewhat or much higher to achieve 80% acceptance, especially of nonsmokers (Walker et al. 1997), than that presented in the EDM's recommendations. Others have concluded, for example, that general dilution ventilation is not recommended for control of odors from many substances (e.g., Knutson 2003). As more research is completed, and practical field experience is gained, updated data will likely appear in publications such as *ASHRAE Transactions*, the *ASHRAE Journal*, and any future revisions or addenda for this book, for example.

4.3.6. Use of Transfer Air in $\dot{V}_{ets}$

An energy conservation opportunity (ECO) refers to a recognized potential for energy savings. Spaces with high ventilation air requirements often present large ECOs. Moving conditioned air from a nonsmoking space to one with ETS via hallways, transoms, grilles, open plenums, or transfer air ducts has long been a technique for conserving energy (Rock and Jadud 1993). The air already used in the nonsmoking space for general ventilation is likely to be much cleaner than the air in an ETS space and that air would eventually be exhausted anyway. Thus, its reuse to ventilate the ETS could be a very good energy conservation measure (ECM). Other ECMs are possible too and will be discussed in the next chapter.

Transfer air through doorways will be discussed further later and is needed to reduce the amount of ETS escaping a smoking-allowed area. But using transfer air to specifically meet the ventilation air requirements of Standard 62.1 may someday become less attractive from an energy-conservation point of view. Neighboring spaces may be required to be over ventilated so that the intended transfer air contains the needed unused air for ventilating the subsequent space. Injecting conditioned ventilation air directly to the ETS area, instead of using transfer air, helps ensure that air is being delivered where needed and may prevent overconditioning of surrounding spaces.

4.3.7. Use of Cleaned Recirculated Air in $\dot{V}_{ets}$

As for the base rates determined via Standard 62.1's Ventilation Rate Procedure, the required ventilation air for ETS areas may be obtained from various sources. Often only outside air is used, but a portion of $\dot{V}_{ets}$ may be obtained via cleaned recirculated air; the data for the volumes of ventilation air (V_{cig}) were obtained with tests that had much recirculated air (Cain et al. 1983, Q&A paper appendix) and thus the air should have been well mixed. Check any codes, standards, ordinances, and other regulations applicable to your particular design job—sometimes only once-through systems are allowed by local ordinances for certain ETS occupancies. But if allowed, using some filtered recirculated air offers the potential for significant energy savings and better humidity control in many climates.

Air from smoking-allowed spaces might be recirculated to those same spaces to help meet the $\dot{V}_{ets}$ needs if the ETS has been cleaned from the air stream. Because filtration in not likely 100% efficient, the flow rate of filtered recirculated air that may be credited toward the ventilation air requirement must be adjusted. The simple relationship is

$$\dot{V}_{ets,f} = E_f \bullet \dot{V}_{ca} \qquad (4.26)$$

where:

E_f is the demonstrated efficiency of the air cleaner for removal of both the gaseous and particulate components of ETS

$\dot{V}_{ca}$ is the volumetric flow rate of recirculated air (Figure 3.1)

$\dot{V}_{ets,f}$ is the resulting flow rate that may be credited

Extreme care is needed in determining the filter efficiency and in specifying construction details and maintenance procedures, as will be discussed further in the next chapter. ASHRAE's SPC 145 is developing a method of test for E_f, and hopefully manufacturers will soon be able to determine and report their values using this proposed standard.

If filtered recirculated air is to be used, and no transfer air, the needed additional rate of outside air ($\dot{V}_{ets,oa}$) is then at least

$$\dot{V}_{ets,oa} = \dot{V}_{ets} - \dot{V}_{ets,f} \qquad (4.27)$$

EXAMPLE 4.2

For the same space described in Example 4.1, a very high efficiency filtration system is specified. The designer is confident, through extensive documentation, that the filtration system will be at least $E_f = 85\%$ at cleaning both ETS's particulates and gases from the recirculated air. In a bold move, it is decided that all of the extra needed ventilation air, 134 CFM, will come from filtering recirculated air. From a rearranged version of equation 4.26, the flow rate of the recirculated air through the ETS filtering system is then

$$\dot{V}_{ca} = \frac{\dot{V}_{ets}}{E_f} = \frac{134\,CFM}{0.85} \approx 158\,CFM \qquad (4.28)$$

The needed HVAC system is then designed and specified. The precise filtration system, necessary replacement consumables, and the maintenance intervals needed should be very clearly documented in the design.

Note that in *no case* should filtered ETS-laden air be recirculated to smoke-free spaces, but filtered air from nonsmoking spaces might be reused in smoking-allowed spaces. Examples will follow in the next two chapters.

4.4. OTHER METHODS

After reading this chapter, it would be incorrect to conclude that the methods presented here—the theoretical rate with perfect mixing method, the ventilation for ETS in aircraft methods, and the ETS dilution method for buildings—are the only ways available for determining ETS ventilation air-flow rates. There are many others, and more will likely be developed in the coming years. Adjustments will likely be made to the EDM too— for example, adding a male/female adjustment factor or possibly increasing the recommended ventilation air-flow rates. As the designer, it is up to you to apply the knowledge and experience you've gained to create the most appropriate HVAC systems needed for your particular applications. The methods you use are limited by only a few things, such as your expertise, responsibility to the clients, and the resources available to you, for example. Should you develop better methods or input data for secondhand smoke design, or make refinements to the existing methods or data, you are encouraged to present your findings to your colleagues through published articles, for example.

5

ETS DESIGN ISSUES

This chapter presents more details and discussion on many issues introduced in the preceding chapters and on new topics as well. The options discussed should help owners, architects, and HVAC designers to find optimal solutions for each, often-unique situation. However, the information is not intended to limit designs to only those presented in this chapter, or elsewhere in this book, but is instead meant to stimulate creative thought.

5.1. METHODS OF CONTAMINANT CONTROL

There are various, basic ways of reducing airborne contaminants, and each has its own inherent advantages and disadvantages. By far the most effective method is *source control*, which means reducing or eliminating the contaminant source or changing the process to something less polluting. *Separation* of functions, such as subdividing a space so that the contaminant is physically limited from spreading to the remainder of the space, is often an effective technique for source control when elimination is not feasible. If either source control or separation is impractical, then the next most effective technique is *local exhaust*; that is, to immediately vent as much of the contaminant as possible before it can mix into the surrounding air. If smokers will likely be at a specific location in a particular space, local exhaust can be effective depending on how well the ETS can be "captured" before it mixes with surrounding air. *Dilution* of the space's air with cleaner air is another method of contaminant control. Because the contaminant has spread throughout the space, to varying degrees, dilution

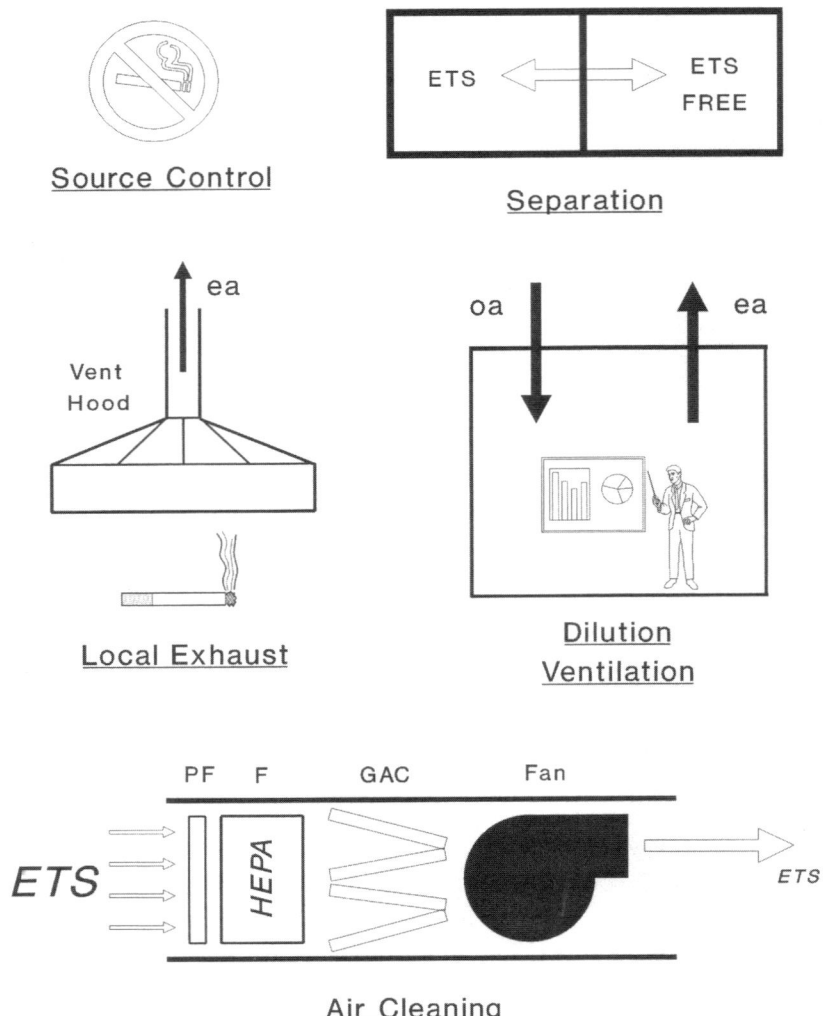

Figure 5.1. Five general approaches to controlling airborne contaminants, including ETS: source control, separation, local exhaust, dilution ventilation, and air cleaning. In the air handler shown the gaseous air cleaner (GAC) removes targeted gases.

is often not the most desirable approach. But it is frequently the most practical. Implied in these first-mentioned methods is that the exhausted contaminants will not be moved to a location near any space's air intake. *Air cleaning* is the final contaminant control approach. All of these techniques, shown in Figure 5.1, and often used in some combination, will be discussed further in this chapter.

5.2. SOURCE CONTROL

The most effective way to minimize the presence of an internally generated contaminant is to reduce or eliminate the source itself. For example, photocopy machines and laser printers produce airborne contaminants as they "fuse" their plastic particles, "toner," onto sheets of paper; these and other machines having open electrical discharges also produce ozone. Recognizing such, it may be possible, for example, to move a high-volume photocopier out of general occupancy space in the center of a building to a small space on the exterior. Through an exhaust fan and ductwork, the machine's fumes can then largely be vented to the outside rather than being allowed to mix with the indoor air. Tobacco smoking source control is achieved through indoor smoking prohibitions.

5.2.1. Design Needs

As we are to design our systems to meet the needs of our clients, we need to communicate with a project's architect and the owners' representatives on ETS and other design issues. As the layout of a building's spaces, form, and functionality is largely in the design domain of the architect, it might not be possible to recommend installing, moving, or eliminating, for example, a smoking area. These restrictions may then require that even more engineering measures be employed to ensure odor and irritant acceptability in these and any surrounding spaces. For example, when a smoking lounge is created, the partitions between it and its surroundings will need especially careful details and construction, and the air-pressure difference may need to be great to reduce the likelihood of significant ETS migration.

If code or ordinance allows smoking indoors, and the client desires such, you are free to suggest that a full or partial smoking ban would have significant benefits, including better IAQ, reduced fire-ignition hazards, maybe lower insurance rates, less maintenance and cleaning, and reduced users' and employees' complaints. But if the owner decides to allow smoking in all or part of a project, you are suggested to record such information and related parameters for use in your design process. Having already accepted the job, it is then your responsibility to design appropriate systems that remove ETS and that reduce airborne migration of ETS to nonsmoking spaces. Obtaining such design expertise through a subcontract to another experienced P.E. or firm, for example, is recommended if not already available in-house. Be sure to consider ETS's complexities and liabilities when setting design fees and acquiring insurance.

5.3. SEPARATION

Assuming that the client wants to allow smoking in at least part of the building, and applicable codes and regulations allow indoor smoking, next the smoking areas need to be defined next. In a general use space, such as a casino, all the floor area may be designated as a smoking area, and separation may not be feasible. But in many other cases, such as in airports, office buildings, and restaurants, it is possible to create separate, designated smoking areas.

An *ETS area* is a space where smoking is to be permitted, as well as any surrounding areas that are not effectively separated. *ETS-free areas* are those where no smoking occurs and the spaces are physically separated from any ETS areas. These definitions, therefore, require that something tangible or engineered must exist between smoking and smoke-free areas, such as partitions and optimized entryways, as will be discussed further. All spaces within a building need to be identified as either ETS areas or ETS-free areas if smoking is allowed anywhere in the building.

5.3.1. Degrees of Separation

The first level of separation, which has been commonly used in the past but is not particularly effective in minimizing ETS exposure, is to simply *subdivide an open space* into smoking and nonsmoking sections, as shown in Figure 5.2. This approach has been popular in existing buildings, as it is easy to do with signage, and so on. For example, restaurants often designate a smoking area so that nonsmokers can be seated outside of it. Unfortunately, many times no efforts have been made to optimize

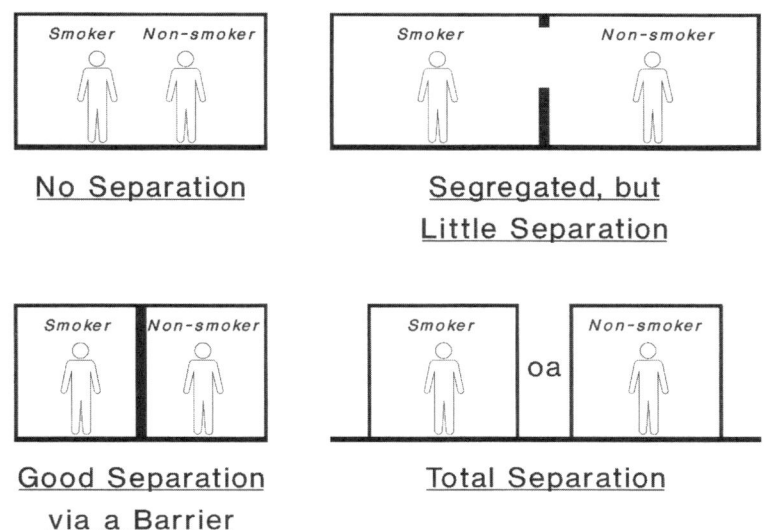

Figure 5.2. The effectiveness of separation measures varies; total separation, via different buildings, is probably the most effective.

the HVAC system when such open-floor separation schemes have been employed. Contaminated air, especially in well-mixed rooms, spreads easily from one section to the other (Cains et al. 2004). Often these spaces have a high-percentage recirculation of the return air, so the ETS is further spread to the nonsmoking sections through their HVAC systems. However, if, as discussed in Chapters 3 and 4, substantial dilution and/or filtration is provided, a majority of occupants may find the air acceptable from an odor and irritant perspective, especially after they've become adapted to it. While many jurisdictions may still allow this approach to separation, stricter requirements are imposed elsewhere.

Physical segregation is the next higher and usually more effective level of separation. The smoking and nonsmoking areas must have effective ETS barriers between them. But the intention of the physical separation can easily be defeated if, for example, a barrier is poorly constructed, a door between spaces is left propped open, or a recirculating HVAC system serves both spaces. A physical, although leaky barrier may increase acceptability somewhat due to occupants' visual perceptions, but such a leaky barrier is not recommended as the sole measure. An even further degree of separation is to make the ETS and ETS-free spaces in *different buildings*, even if only separated by a few feet of outdoor air. When codes or ordinances require physical separation, they also typically call for separate ventilation systems for the ETS and ETS-free spaces. These separate systems must not transfer air from ETS areas to ETS-free spaces.

The last degree of separation is to restrict either smokers or nonsmokers from a building. Typically this is accomplished via an *indoor smoking bar*. In some cases, such as for a private club or residence, the full building may be declared by the owner/operator as an ETS space via signs posted at all entrances or by other methods, and nonsmokers must then choose whether or not to enter.

5.3.1.1. Barriers

Walls, floors, ceilings, and other space-dividing partitions that separate an ETS area from other parts of a building need to be effective at reducing the transfer of ETS from one side to the other. While materials and/or construction assemblies are not perfect, good choices and subsequent high-quality construction techniques can minimize ETS movement. As will be discussed later, negative air pressurization will help to further reduce the migration of ETS across these *barriers*. Walls and ceilings to be constructed as these smoke barriers should have proven air retarders, especially where negative pressurization is not possible. One common type of *air retarder*, commonly called a "vapor barrier," is polyethylene sheeting, but even a well-caulked and oil-painted wall can be relatively effective at minimizing air leakage. Special care is needed to seal electrical boxes and other surface penetrations, via caulking and gasketing for example, in any smoke barrier as shown in Figure 5.3. Air leakage through buildings is discussed in more detail in the "Ventilation and Infiltration" chapter of the *ASHRAE Handbook* (ASHRAE 2001) and elsewhere.

5.3.1.2. Connecting Doors

Unless an engineered, one-way airflow entryway is designed, as discussed later, doors are needed to define the separation of the ETS area from its surroundings, to allow passage of occupants, and to help keep ETS inside. Some common types of doors are shown in Figure 5.4. Research (e.g., Alevantis et al. 2003) and experience have shown that swinging and rotating doors are not optimal choices for ETS applications, because signifi-

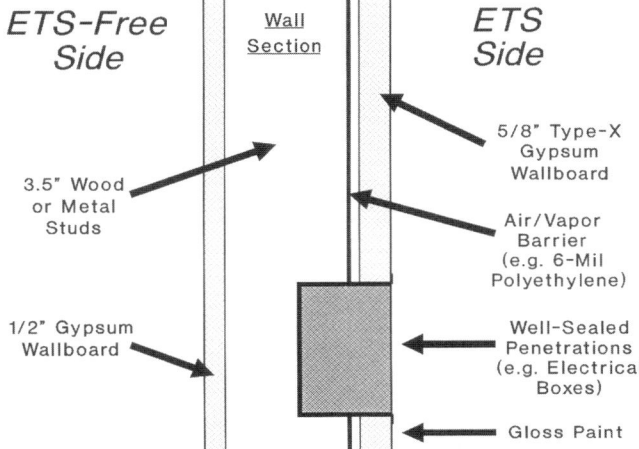

Figure 5.3. A barrier must substantially prevent the migration of ETS through a wall, ceiling, or floor, for example. Shown is one possible detail for a lightweight interior barrier wall.

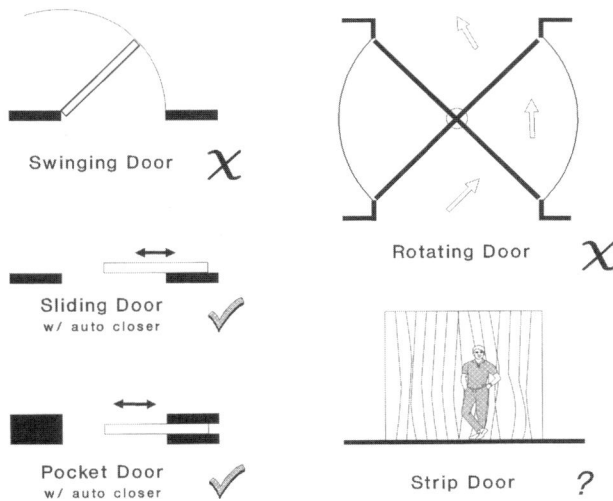

Figure 5.4. Swinging and rotating doors should not be installed between ETS and non-ETS spaces; instead, sliding or pocket doors with automatic closers should be used. Industrial-type strip doors may or may not reduce the pumping of air between the spaces to an acceptable level.

cant volumes of air are "pumped" from one side to the other with each use. *Sliding* and *pocket doors* dramatically reduce this quantity of air transferred, but they can't eliminate it altogether. Industrial strip-type doors may also have somewhat lower air-transfer volumes than swinging or rotating doors but are not likely to be acceptable to owners and occupants in most commercial applications.

It is logical that doors between ETS and ETS-free areas should have *automatic closers* to minimize the potential for doors being left open. These could be simple spring/hydraulic devices or more complex electrical/mechanical closers. A combination of a double-sided (inside and outside) motion detector and a motor-driven door, with minimal time-open delay, might be optimal where the budget allows for such. Additional guidance is needed on specifying the delay, fire-egress mode, and so on, so consulting the manufacturers' literature and their sales representatives, for example, is advised. For reduced air transfer, a *vestibule* or "air lock" can be created by having two or more doors in series, as shown in Figure 5.5 The doors need adequate separation to allow safe passage, and the space between the doors should be at an air pressure between that of the

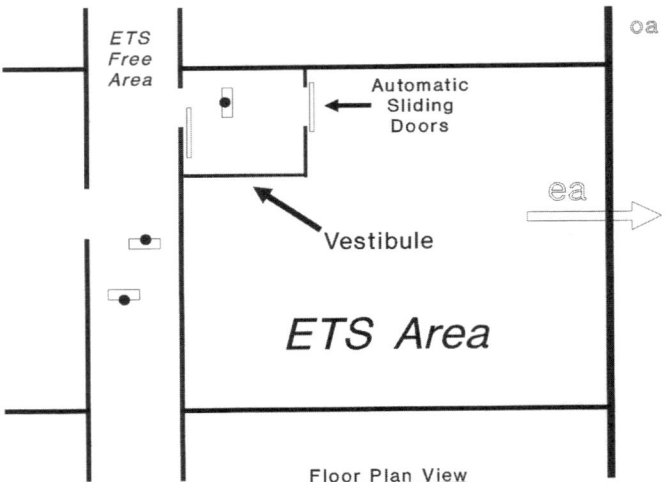

Figure 5.5. A vestibule is a space created between two or more layers of doors. It is typically used for exterior entrances to reduce infiltration but can be used indoors to decrease ETS and other airborne contaminants' movement.

ETS and ETS-free areas to ensure that the dominant air-flow direction is into the ETS space.

An alternative to using an automatic door closer is an engineered system that similarly reduces passage of ETS. One such alternate approach, setting an appropriate, continuous air-flow rate through an opening, will be discussed later in this chapter.

At first glance, *air curtains*, which are vertically projected jets of air over a doorway (ASHRAE 2004, ch. 17), may seem attractive for ETS applications. Occupants could easily move from space to space, and a smoker could be "air washed" before he or she leaves an ETS area to slightly reduce carried-through odors. Air curtains may reduce flying insect movement across them, but users might find their air blasts objectionable. However, great care is needed if air curtains are considered. Because an air jet always entrains surrounding air, air from both sides of the doorway will be mixed together, as shown in Figure 5.6. As the ever-growing air jet reaches a boundary, such as the floor, it then will spread horizontally. Large quantities of mixed air are then likely to be injected into the rooms on both sides of the doorway. This means that much ETS will likely be transferred to the neighboring ETS-free space, which would be inappropriate. There could be, of course, exceptions via advanced designs

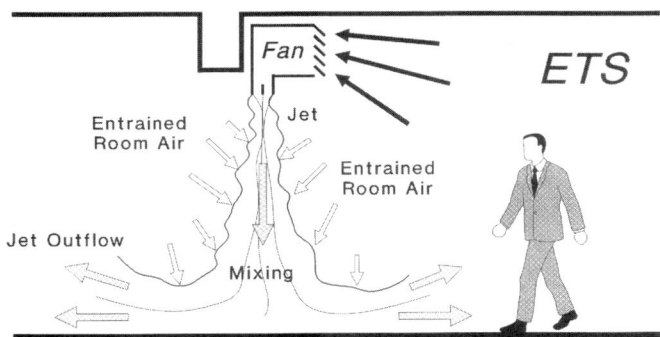

Figure 5.6. Air curtains are often used on doorless entries to reduce heat loss somewhat. But because air jets have inherent entrainment and mixing, ETS will move from one side to the other, potentially in great quantities. Aiming the jet inward somewhat reduces the outflow but does not eliminate it. Drawing air for the curtain from the outdoors or a non-ETS space should also decrease the ETS transfer across the curtain.

that minimize air transfer out (Hyvärinen et al. 2002), but such should be examined carefully and likely proven first via lab and field tests.

Other barrier penetrations between ETS and ETS-free areas need to be avoided where feasible, and otherwise minimized. Holes drilled for the passage of conduit or piping, for example, should be only slightly larger than the objects themselves, and then the remaining gaps should be sealed. A fire-resistant or intumescent sealing material may be advisable, or even required by local fire code or fire marshal directives. Passages for ducts or structural elements should be sealed as well. Access hatches and the like should be kept closed and be well-gasketed to minimize air leakage. While it might not be required by local regulations, increasing somewhat the fire rating and other safety features of an ETS area may be wise. When laying out a smoking area, be sure to comply with the other fire and life safety requirements too; for example, two means of egress are often necessary, so a door to the exterior may be needed, and automatic sprinklers may be required (e.g., NFPA 1, 13, 101; various years; local ordinances).

Design, construction, and operation vary, so the actual leakage rates of ducts, barriers, or doors are unknown. For critical applications physical testing of the as-built rooms or assemblies may be required to verify their predicted behavior. As discussed later in this chapter, providing means for adjusting systems after their installation and testing is important for achieving the systems' optimal performance.

5.3.1.3. Outdoor Smoking Areas

Whether indoor smoking areas or bans are included or not, designating outdoor smoking areas is advisable. If not, people are likely to smoke outdoors in locations that are undesirable, such as near air intakes or operable windows; in or around locations of high fire hazard, such as fuel meters or storage tanks; or at the front entrance. In dense urban areas, outdoor smoking areas may not be possible except on the buildings' roofs or balconies, for example, and again care is needed that smoking is not done near air intakes or operable windows. The public should not have to walk through the smoking area for entrance, egress, or to access restrooms, for example. Nor should waiting areas be smoking allowed, because non-smokers would be present.

When possible, outdoor smoking areas should be covered to help protect users from the elements, but they also should be designed so that air can easily flow through the area to dilute and carry away smoke. Outdoor smoking areas should not be too near building walls when air intakes or operable windows are situated nearby or above; the "boundary layer" and "separation regions" (ASHRAE 2001, ch. 16) will keep the smoke concentrations high near the building when weather conditions are favorable. A competing concern is that smoking spaces should be close to workstations so that breaks can be as short as possible. Having several smaller, distributed smoking areas, indoors and/or out, may thus be more effective than designing one large centralized space.

5.3.1.4. Smoking Receptacles

Ashtrays and urns, or any other repository for used and possibly still smoldering tobacco products, need to be strategically placed to maximize their use. But they also should be well away from ETS-free areas, air intakes, and operable windows, for example. Outside, they should be placed so that smokers entering a building can extinguish their products; but again, they should not be so close to the doors that nonsmokers will be highly exposed to any smoldering product or last exhaled smoke. Some repositories include water-baths or other techniques to extinguish leftover product, but the required maintenance of such devices may be impractical for some applications.

Within indoor smoking-allowed spaces, placing ashtrays at seating areas near exhaust grilles can encourage users to stay nearby, which increases the effectiveness of the local exhaust measures and should help remove smoke from any leftover smoldering products. Placing a large urn near the exit is a common practice, but locating such farther away from the exit, but not too far, encourages users to extinguish their product and hopefully to exhale a time or two before leaving the room; ETS annoyance complaints can be generated by smokers even when they are not smoking.

5.4. LOCAL EXHAUST AND AIR MAKEUP

After source elimination and/or separation, capturing the contaminant as close as possible to the source, and then exhausting it to the outdoors, is the next most effective method of control. Unfortunately, smokers can be spread throughout a room, and they often move about while smoking. But in some applications you can predict locations where smokers will most likely be residing. For example, in a restaurant, most smoking customers will be seated, so having the architect fix locations of the tables in advance can help you to place provisions for local exhaust. An exhaust *hood*, with or without an integral light fixture, could be specified over such tables, similar to that shown in Figure 5.7. "Capture" of buoyant pollutants can be increased by placing a hood above and as close to the source as is reasonably possible and by not having supply air jets crossing the immediate area. The hood should be broad.

If the specific locations for hoods won't be known until construction is relatively complete, specifying adjustable exhaust ductwork can allow for some fine-tuning as opening day nears. More specific exhaust examples will be given in the next chapter, and many more details on hoods can be found in the *ASHRAE Handbook* (ASHRAE 2003b, ch. 30) and via ACGIH (2001).

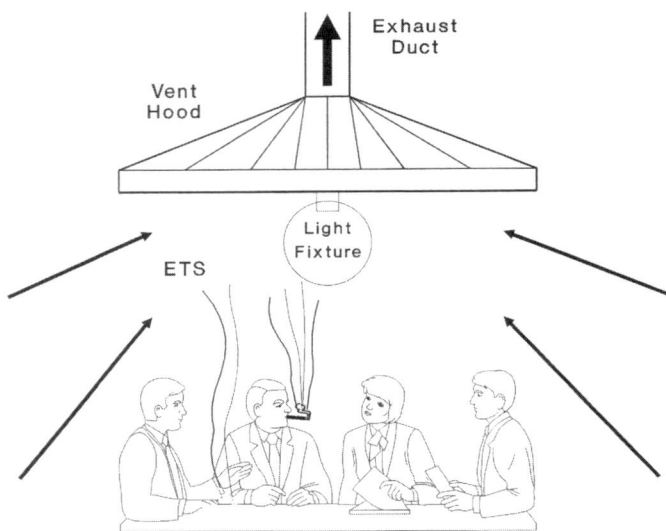

Figure 5.7. Local exhaust via vent hoods over known locations of smokers can be more effective than general dilution alone. Hoods should be wide and also placed as close as possible above the smokers. They won't catch all the smoke, so general ventilation, exhaust, and/or air cleaning will be needed too, as will considerable makeup or transfer air.

5.4.1. Exhaust Fan Placement

If an exhaust duct from an ETS area passes through other parts of a building, there is potential for leakage of ETS-laden air into the surrounding spaces. Operating such exhaust ducts at a negative pressure is highly recommended and may be required (e.g., Standard 62 addendum *v*, ANSI/ASHRAE 2001; Uniform Mechanical Code [UMC]). Specifying high-quality, durable ducts and duct sealing can reduce this air leakage, and thus also increase air-flow rates from the intended locations. Figure 5.8 shows that mounting the exhaust fan at the end of the duct run can achieve the relative negative gauge pressure in the exhaust ductwork. If mounted on a flat, walkable roof, or within ladder height on an exterior wall, inspection and maintenance of such fans is relatively easy.

As high exhaust air-flow rates are likely required to remove the internally generated ETS, similar amounts of conditioned makeup air will be needed to replace it. As discussed previously, a popular source of makeup air is air transferred from surrounding spaces. Then those other spaces will need their own makeup air. Introducing supply air directly to the ETS

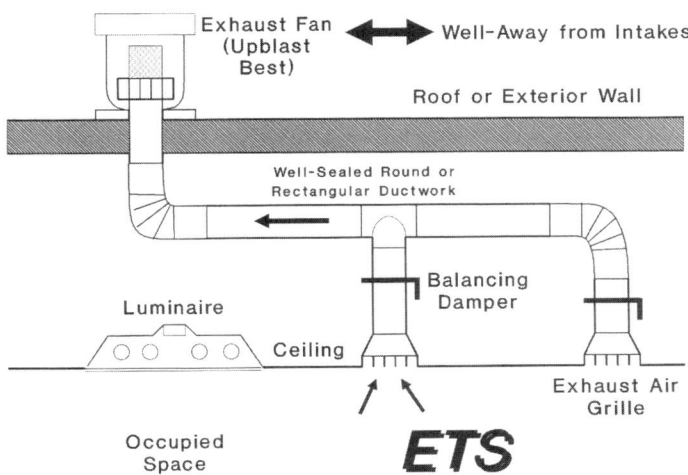

Figure 5.8. Exhaust fans should be placed at the end of the duct. The highly negative pressure created in the ductwork should prevent leakage of ETS into the building. The duct should be well sealed to ensure that the needed exhaust air-flow rate from the ETS space is achieved.

space as the makeup air is an option too. A combination is possible as well. Varying the supply, transfer, and exhaust air flows can be used to maintain a desired pressurization, as will be discussed later in this chapter.

5.4.2. Separation of Intakes and Exhausts

With either localized or general exhaust of ETS-laden air, reentrainment of the diluted ETS back into the same or different building should be minimized. How to design stack heights for optimal dispersion of industrial products is fairly well known, but research is ongoing for more basic air exhausts from buildings. The relatively new "Building Air Intake and Exhaust Design" chapter of the *ASHRAE Handbook* (ASHRAE 2003b) is being expanded to provide guidance, but it is not yet as practical as some designers desire. It and the "Airflow Around Buildings" chapter of the *Fundamentals* volume (ASHRAE 2001) do describe the basic physics of plume rise and dilution, but simple yet highly accurate methods for predicting actual minimum dilution ratios, which are dependent on many factors, including the weather, will probably not become available for everyday design purposes. But various attempts are being made and, when available, are being brought into the *Handbook* and Standard 62.1, so more guidance on this issue should be available with time. One practical but not highly accurate approach relies on the "stretched-string" (S)

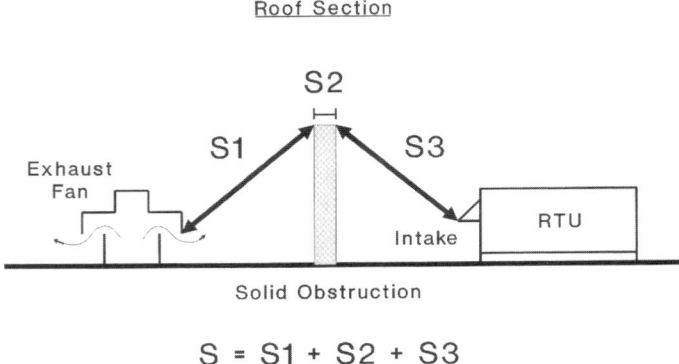

$$S = S1 + S2 + S3$$

Figure 5.9. Outdoors, exhausts should be placed as far as practical from all air intakes. The "stretched-string distance" (S) is the separation, and a minimum value may be mandated by code (Rock and Moylan 1998).

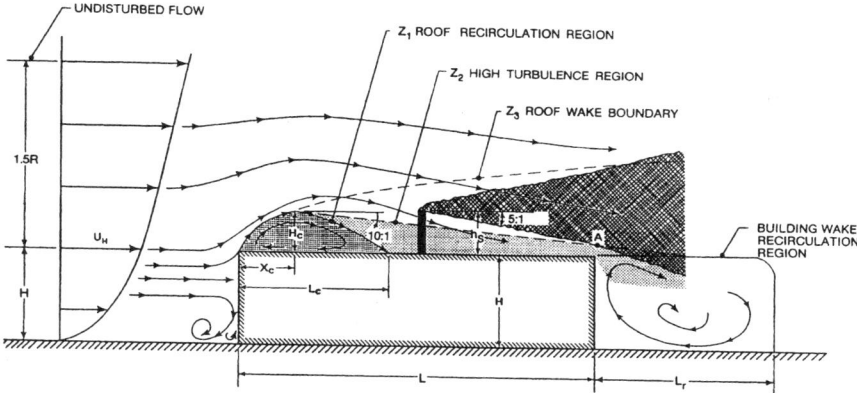

Figure 5.10. Some ETS exhausts may need to be designed similar to industrial stacks. Much information is available on dilution and plume spread in the *ASHRAE Handbook* and elsewhere (e.g., Wilson 1979; ASHRAE 1997, ch. 5.12).

distance between the exhaust and intake, as shown in Figure 5.9. Codes, for example, can give minimum required stretched-string distances that hopefully will achieve good dilution for most cases.

When necessary, exhaust ducts and fans for ETS-laden air can be designed using plume rise and dispersion, similar to that done for laboratory fume hoods, to further minimize reentrainment. Use as tall a stack as

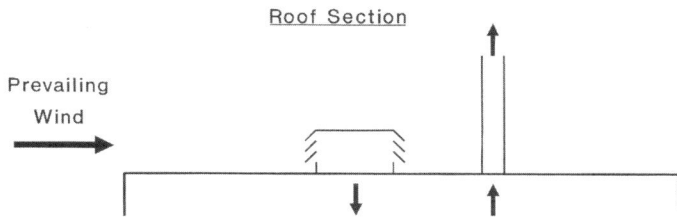

Figure 5.11. When a dominant wind direction exists and is known, the intake should be upstream of the exhaust (Rock and Moylan 1998). But some experts argue that there typically are no "prevailing wind directions."

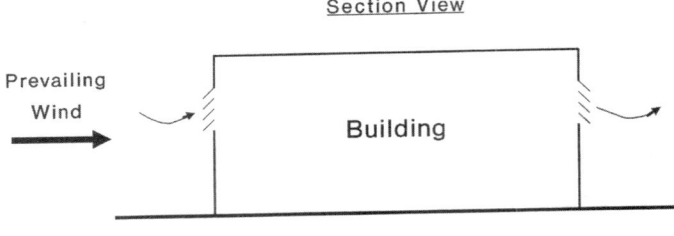

Figure 5.12. Wall-mounted intakes and exhausts should be as far above the ground as practical, and the intakes should be upstream (Rock and Moylan 1998).

feasible, and impart a high air velocity so that the resulting plume travels far vertically and entrains more surrounding air to enhance dilution (ASHRAE 2003b, ch. 44). Figure 5.10 shows such an arrangement. Figures 5.11 to 5.13, from an ASHRAE research project report, show some suggestions for locating general building exhausts relative to air intakes when traditional exhausts are used.

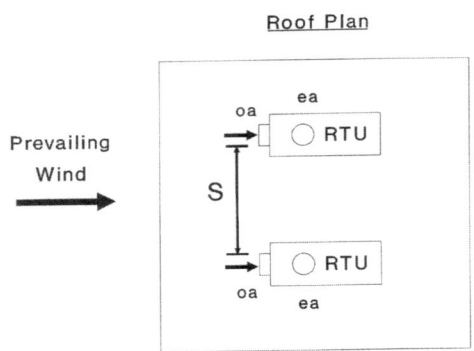

Figure 5.13. When rooftop units are used for venting ETS, they should be as far as possible from each other. Outside air intakes should be closest to the prevailing wind direction (Rock and Moylan 1998).

5.5. DILUTION WITH OUTSIDE AIR

Because of the mobility of smokers within spaces and air mixing, at least some degree of dilution will always be needed to help ventilate an ETS space. In many cases, dilution may be the only feasible control choice. The bulk of this book, and especially Chapter 4, assumes dilution ventilation, but you should be sure to consider other and possibly more effective approaches to removing ETS, such as displacement ventilation or local exhaust.

5.5.1. Displacement Flow Helps

Careful selection of diffusers can yield air mixing performance, and thus dilution, similar to that of perfect mixing. However, creative placement of air outlets and inlets can make, at least somewhat, a degree of advantageous displacement flow. Rock and Zhu (2002) and the *ASHRAE Handbook* (ASHRAE 2004, ch. 17) provide detailed information on types of diffusers and grilles, and the selection process. Figure 5.14 shows that by placing outlets low and exhausts high, ETS-laden air can be encouraged to move more rapidly from the occupied zone of the room to the ceiling

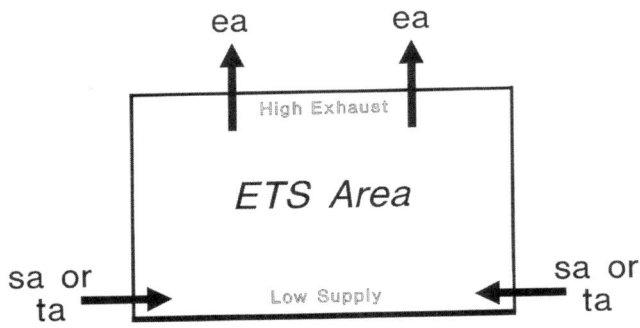

Section View

Figure 5.14. Injecting the supply or transfer air low helps ventilate the occupied zone, or lower 6 ft (1.8 m), of a space. Allowing the ETS to rise by buoyancy and then be removed from the ceiling level also helps.

area. There still will be mixing of the air and exposure to ETS in the occupied zone but hopefully to a reduced degree. If horizontally supplied transfer air is to be used, such as that through an open entryway, a degree of side-to-side displacement flow is created, but downstream occupants may experience higher exposures than if the room were perfectly mixed. As also discussed in Chapter 3, once-through floor-to-ceiling displacement flow seems optimal for ETS areas when possible, because the warm, buoyant air and contaminants from occupants and tobacco combustion will naturally want to rise anyway.

5.5.2. Jets

As with doorway air curtains, *air outlets* should be used near doorways in ETS spaces only, if at all, with great care. For example, a nearby four-way blow ceiling diffuser will produce a jet of air in the direction of the doorway. The jet's momentum may overcome the negative pressure differential across the doorway and send air, composed of both the jet's primary air and entrained ETS-laden room air, out an opened door or a doorless entryway. Most of the exhaust air should be drawn from other parts of the ETS area so that the dominant in-room air-flow direction, as much as is possible, will be away from the doorway and back into the room, as shown in Figure 5.15. To help this effect, it might be possible to

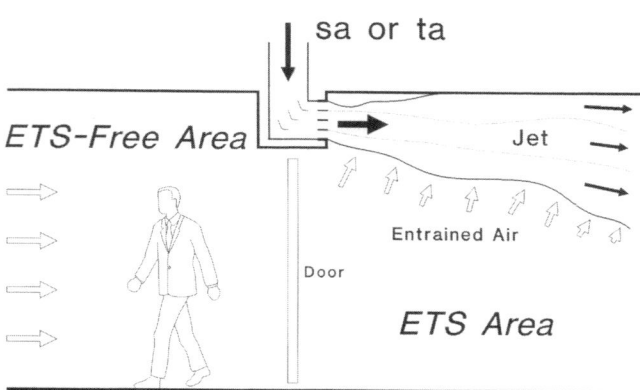

Figure 5.15. Do not blow air toward the exit from an ETS space. But having a jet blowing back into the room, or an exhaust above the exit, may help reduce the amount of ETS leaving the space.

select a high-induction, one-way air outlet and place it just inside the room above the entryway; this high-sidewall supply air grille, for example, would entrain ETS-laden air from below and blow it back into the room and toward exhaust grilles, thus reducing the ETS migration out of the entryway.

In general, *ceiling* and other in-room *circulating fans* should not be installed or used in ETS areas. As with diffusers placed too close to entryways, fans can increase air transfer out of spaces. And while in-room fans can be useful in increasing thermal comfort, they also mix the buoyant ETS plumes into the room air; instead, the plumes should be allowed to rise above the room's occupied zone and from there be exhausted outdoors, or potentially filtered from the air.

5.6. CLEANING AIR

Another option for control of airborne contaminants, both gaseous and particulate, is air cleaning. Some air cleaning is required in every air-based HVAC system to at least protect the equipment. As described in the previous chapter, effective air cleaning can be used to provide some of the ventilation air needed for ETS areas if allowed by code or ordinance but should be applied with great care. However, logic requires that air may not intentionally be transferred or recirculated from an ETS area to an ETS-free area. For example, ANSI/ASHRAE Standard 62.1-2004, Section 6.2.9, states that "air from smoking areas shall not be recirculated or transferred to no-smoking areas." Implicit in this requirement was that cleaned ETS air may not be moved to an ETS-free area, even if 100% cleaning effectiveness is claimed, but this does not preclude using such cleaned air in the same or another ETS area. If such an approach is used, as shown in Figure 5.16, moving the filtered air in the direction of increasing ETS concentrations is probably wise.

As shown in Figure 3.1 by the position of the filter (F), in simple recirculating systems, air treatment is most frequently done on the mixed air, just before the coils. But for specialized treatment, such as for reducing sulfur compounds in the ambient, admitted outside air, putting a treatment device in that particular air stream may be most effective; the concentration of concern is likely higher there, and the air-flow rate is lower than after mixing with other air streams. The contaminants of concern could be particles or gases, or, in the case of ETS, both, so more than one cleaning technology must be applied.

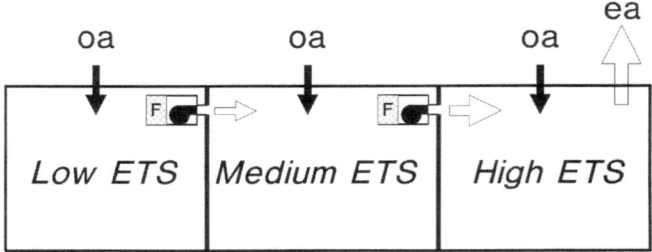

Figure 5.16. Cleaned or other air should not be recirculated from a smoking area to a nonsmoking area. However, the cleaned air can be recirculated to the same space or other ETS areas. Where there are expected gradients of ETS from space to space, it may be advantageous to "leap-frog" the cleaned air toward the eventual exhaust from the most contaminated space.

5.6.1. Removing Particles

Air in and around buildings always contains particles in a wide variety of sizes from sub-0.1 μm, for some components of smoke, to greater than 100 μm for large allergens; air cleaners need to be matched to the need. As this range is over many orders of magnitude, and the mixtures and concentrations vary significantly in any particular space over time, it is very difficult to select one specific air cleaner that will perform at all times for many different applications (ASHRAE 2004, ch. 24). Building owners and operators should expect that some adjustments in the applied filtration technologies will be needed to optimize their performance and costs. Such changes should be done in consultation with a ventilation system design engineer.

Most *air filters* for HVAC applications are disposable, but a few may be cleanable. While the performance of new, high-quality replacement filters that meet the desired standards is relatively assured, care is needed by owners/operators so that any cleanable filters are eventually replaced, because they do deteriorate with repeated use. The highest-performance filters often are available only as disposable units. Filters need to fit tightly into their *filter frames*, have good seals between filters and frames, and have gasketing on access doors to minimize air bypass.

The *filter efficiency*, E_f, is a filter's ability to remove particles from a given airflow when tested with an approved "challenge aerosol." As the particulates vary greatly in size, and the filter efficiency alone does not fully describe its ability to remove all sizes of dust in a real application, E_f should be used with care. The *dust-holding capacity* is a filter's ability to retain the gathered particles and is directly related to the needed frequency of replacement. Associated to both the efficiency and capacity is the filter's *resistance to airflow* (ΔP_f), which is the pressure drop across it at a given air velocity. The ΔP_fs will be greater for higher-efficiency filters, and also increase as filters "load up" with use. A system's fan needs to overcome this and other pressure losses, so a higher resistance to airflow implies more fan power and increased energy consumption. For ETS applications, initial pressure losses through the filters of 0.5 to 1.0 in.w.g. (124 to 249 Pa) or so are common, where for general ventilation 0.2 to 0.3 in.w.g. (50 to 74 Pa) are more typical. Installing "filter service" differential pressure sensors across filters can encourage replacement of filters when needed; the "initial" and "final" pressure losses are highly dependent on the specific filtration used. For ETS applications, due to odor build-up, filters may need to be changed even more often than the pressure drops across them indicate.

5.6.1.1. Filter Ratings for ETS

There are many designs of air cleaners, so classifying them and then evaluating and reporting their efficiencies is difficult. Performance tests have evolved over time and are reported on in the *ASHRAE Handbook*, standards, and elsewhere. One method of test (MOT), provided in ASHRAE Standard 52.2-1999 for media filters, yields a minimum efficiency reporting value (MERV) from 1 to 20, which describes a media air cleaner's performance at removing particles in the range of 0.3 to 10 μm. ASHRAE Standard 52.1-1992, which is being phased-out in favor of 52.2, includes the "dust spot efficiency" and defines the filter's *arrestance* as its ability to stop some standardized dust; this standard is now to be used for low-efficiency (~MERV 1-4) filters. Table 3 of ASHRAE Standard 52.2, and repeated in the *Handbook* (ASHRAE 2004, table 3, ch. 24), recommends that filters with a MERV number of 15 or higher be specified for ETS applications. The filters are usually very deep, often 12 in. (30.5 cm) or more in thickness, so considerable space is needed for them, their frames, maintenance access, and for dry storage of replacements. They usually won't fit in common RTUs and fan-coil units, for example, so larger or custom units may be required.

Another media filter performance test is the military's DOP Penetration Test, which measures the degree of passage of a special vaporized liquid, dioctylphthalate (DOP). The test's "challenge droplets" of interest are very small, approximately 0.3 μm, and are similar in size to the smaller particles in ETS. When available, typically for MERV 16 or higher filters, DOP-rated filters may be useful for ETS design where extremely high RSP removal rates are desired.

5.6.1.2. Types of Particle Filters

The five main categories of particulate filters are *panel filters*, *pleated* and *pocket filters*, *dry-type extended-surface filters*, *renewable media filters*, and *electronic air cleaners*. A common example of a panel filter is the typical, low-cost residential 1 in. (2.54 cm)–thick "furnace filter"—it has only minimal filtering capability (MERV 1 to 4, typically) and is installed to protect the equipment from large particles. As such, low-MERV panel filters alone would be highly inappropriate for use in recirculating ETS applications. However, Figure 5.17 shows that panel filters *are* often used as part of a larger filtration system; the panel filter serves as a low-cost, frequently replaced prefilter that may noticeably extend the life of the

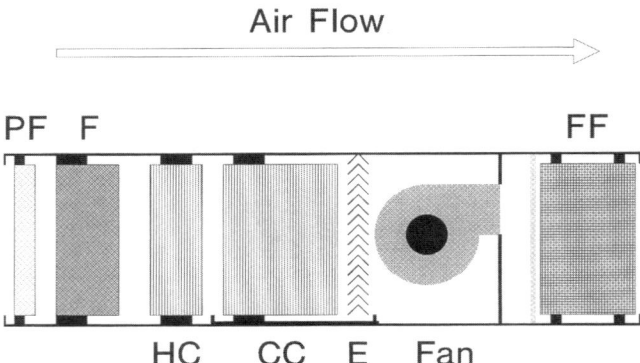

Figure 5.17. A sole air filter may be employed, or an entire filtering system may be used instead. Shown is an air-handling unit that incorporates a prefilter (PF), filter (F), and final filter (FF) for particulates, from low to high efficiency, to achieve very clean supply air. The prefilter may be a simple pleated unit that is changed often.

downstream, expensive high-efficiency filters. A few people, however, recommend against the use of prefilters due to the added pressure loss and cost. Besides the prefilter and filter, some critical applications require a "final filter," sometimes placed after the fan and coils, with the efficiencies of the filters increasing the further downstream each is located.

Most commercial air handlers use filters that have folded media in them, and these filters tend to be 2 in. (5.1 cm) or more thick and of MERV 5–8. The media is folded or "pleated" in these filters to give them more surface area so that the air-flow velocity per unit area of media is lower than for a panel filter. The increased media area also provides more dust-holding capacity. The typical throw-away, MERV 5–8, 2–4 in. (5.1–10.2 cm) -thick pleated filters that we often specify for general applications are not suitable for removing ETS. "Bag," "pocket," or "cartridge" filters are often up to 3 ft (0.91 m) deep for increased capacity; they are usually MERV 9–12 and thus are also not suitable for use with ETS alone. These filters can be used as prefilters for higher-efficiency units downstream, however.

"Renewable media filters," often in the form of "automatic roll" or "moving-curtain" filters, allow the used media to be replaced with fresh while the system is still in operation. A differential pressure sensor or timer, an actuator motor, and a controls system can automatically advance the filter media when a preset condition is reached. As the efficiencies of such filters are typically only 20% or so, about MERV 4, they should not be specified as final filters in ETS applications. They may, however, be useful as prefilters to extend the maintenance interval of the final filter.

"Extended surface filters" are also available with much higher MERVs. *High-efficiency particulate air* (*HEPA*) filters have very dense media and have DOP efficiencies of 99.97% or higher. These HEPA filters should be specified for ETS applications for RSPs' removal. If even higher, clean-room-like performance is desired, *ultra low penetration air* (*ULPA*) and *super ultra low penetration air* (*SULPA*) filters are available, but these are even more costly and can have final resistances of 2.0 in.w.g. (498 Pa) or so (ASHRAE 2004). ULPA and SULPA filters are considered "excessive" for most ETS applications; HEPAs are recommended. HEPA filters are disposable by common methods and are often sold as "rigid cell" or "cartridge" filters. Pleat prefilters of MERV 5–8 are often installed before HEPA filters. Figure 5.18 shows several common types of disposable filters.

Electrostatic precipitators, normally called *electronic air cleaners* (*EACs*) when used in HVAC systems, can be very effective at removing

Figure 5.18. A 1 in. (2.54 cm)–thick panel filter ("furnace filter"), 2 in. (5.1 cm) pleated filter, and 12 in. (30.5 cm) cartridge filter. Having deeper and more pleated media normally implies higher removal efficiencies and capacity but not always. HEPA final filters are normally used in ETS applications.

the finest particles present in ETS. There are many different designs. However, they clog and short-circuit easily on accumulations of larger particles, so a quality prefilter is essential with their use. Electronic air cleaners work by ionizing incoming airborne particles via two high-voltage potentials. The charged particles are then attracted to and collect on downstream surfaces, called the "collector." Periodic disconnection and washing of the collector is needed and can be a frequent, significant, and somewhat challenging task as the electrical elements tend to be delicate. For very large applications, consider specifying electronic air cleaners that are self-washing, which should greatly extend the periods between needed manual cleanings.

In-room floor, wall, or ceiling mount HEPA and/or EAC units, with integral fans, are available. Manufacturers' recommendations must be consulted but a rule of thumb is that they need to provide at least 10 air changes per hour for ETS and similar applications. These units often also include provisions for removing vapors as well as particles.

5.6.2. Reducing ETS's Gas-Phase Contaminants

Gases are generally more difficult to remove from air than particles, and, as such, the devices available tend to be more expensive, require more frequent maintenance, and often have short useful lives. Their performance, however, makes them very useful in ETS applications. Disposable pleated gas-phase filters are available, as are other forms such as ring-panel and standard granular media; when loose absorptive media are used, pellets or flakes (Bohanon et al. 1998) are usually placed in perforated trays (ASHRAE 2003b, ch. 45). These granular media are typically used only for very large commercial and industrial applications. When media reach *saturation*, they should be replaced or renewed.

Other processes are available, such as wet chemical scrubbing or combustion of the contaminants, but these are extremely expensive; most ETS applications rely on readily available "dry scrubbing" (Muller and England 1995) *activated carbon* and/or *potassium permanganate* filters (Liu et al. 1991; Muller and Henriksson 2000). These materials have a high capacity for adsorbing odors from smoke (ASHRAE 2003b, table 7, ch. 45), and are often used in combination with HEPA filters for recirculating ETS applications. They are placed downstream of the HEPA filters.

The sorptive capacity of the gas-phase filters is normally expressed as a percentage of their weight—for example, 20% to 40%—so the frequency of changing can be estimated using the rate of entering contaminant versus the amount of activated carbon in the filter unit. But because actual conditions will vary, the filter service intervals for both the gaseous and the particle filters will likely be developed over time from the maintenance staff's field observations of pressure drops and odors. An ASHRAE standard test method, via proposed Standard 145P for gas removal effectivenesses and capacities, is in preparation; until available, designers must examine manufacturers' information carefully. For example, the recommended minimum residence time in the media may be 0.12 seconds or so for achieving optimal effectiveness.

5.6.3. Sizing and Selecting Filters

The performance of filters, and their air-flow requirements, varies significantly, so manufacturers' literature and air filter representatives should be consulted for their specific recommendations. The National Air Filtration Association (NAFA) is also a good source of information (NAFA 1996). Most particulate and gaseous filters desire a relatively low *face velocity* for the air entering them—for example, 50 to 250 FPM (0.25 to 1.25 m/s)—while air handlers and coils themselves are often sized for 500 to 700 FPM (2.5 to 3.6 m/s). For example, a 2,000-CFM air handler is often about 2,000 CFM/500 FPM = 4 ft^2 in cross-section, but 2,000 CFM/ 250 FPM = 8 ft^2 of MERV 5–8 filters may be required for just normal (non-ETS) occupancy. Panel and pleated filters—for example, 2 in. (5.1 cm) deep—are frequently installed in multiple, slanted frames to increase their numbers, as was shown in Figure 5.1. The various extended surface filters often allow higher face velocities of up to 500 FPM (2.5 m/s).

The example calculation toward the end of the previous chapter showed how to find the flow rate through an ETS-removing filter to provide some of the ventilation air. When possible, consider placing this ETS filter in the recirculated air stream, possibly with a supplemental fan to overcome the extra pressure drop. This minimizes the air-flow rate through the filter and exposes it to the highest concentration of ETS to maximize the removal effectiveness. An additional, lower-efficiency (e.g., MERV 5–8) filter will be needed to protect the coils from outside air dust and could be installed either in mixed air or, if the ducts are well sealed, in only the outside air stream.

The example given in the previous chapter assumed that the space had at or near-perfect mixing. If not so, the capacity of the ETS filter will need to be adjusted. Liu et al. (1991) and others have studied this problem and proposed various calculation methods. ANSI/ASHRAE Standard 62.1-2004 provides detailed equations for making this adjustment. All equations require knowing the ventilation effectiveness, which, as previously discussed, is difficult to predict or measure, so a conservative value should be used from the table provided in the standard.

For general ventilation, using cleaned recirculated air to provide some of the ventilation air requires that you use the IAQ Procedure of Standard 62 (Interpretation 62-2001-17); you will need to meet the procedure's testing and documentation requirements. Debate on the more recent Standard 62.1-2004 seems to have reached consensus that the IAQ Procedure *can't* be used with ETS, but something more than the Ventilation Rate

Procedure's flow rates are needed when ETS is present. Either way, it is in your and your clients' interests to be involved in the postconstruction evaluation of the system to be sure that the desired ventilation performance is being achieved. Also, use the opportunity to critique your and others' systems and details, and use these field observations to improve your future designs.

5.7. PRESSURIZATION

Pressurization, and/or *depressurization*, is usually an effective method for controlling the direction of transfer air and contaminant movement between spaces. When an air-pressure differential exists, air will attempt to move from the higher pressure region to the lower. The air will travel through all available openings, whether they be large, such as open stair- or entryways, or small, such as cracks around electrical boxes. The momentum in high-velocity air can overcome a pressure differential, however, so air should not be blown against openings in ETS barriers. Vapor-pressure or concentration differentials cause diffusion of substances; when setting air-pressure differentials, this additional "driving force" should be considered as well.

By depressurizing an ETS area relative to its neighboring spaces, via exhaust, air should transfer to the ETS space, rather than having significant quantities of ETS-laden air escape. Alternatively, all the surrounding spaces can be highly pressurized, thus forcing air into the ETS area, but this often is not as practical as just exhausting air from the smoking space. Overall, buildings are often operated at slightly positive pressures relative to the outdoors to minimize infiltration. Concerns about condensation in buildings' envelopes are causing many designers to seek near-neutral pressurizations instead.

Logic suggests that ETS-free areas be positively-pressurized relative to a neighboring ETS area, whether separated by a partition or located down a connecting hallway, for example. This does not mean that either type of space must be +P or –P relative to the outside, but only that

$$\Delta P = P_{ETS\text{-}free} - P_{ETS} > 0 \qquad (5.1)$$

This equation does not prescribe a specific ΔP, only that ETS should not migrate from the ETS area to the ETS-free space. Significant exceptions exist for multifamily housing and other uses and reflect the problem of transient rental occupancies, ownership rights when dwellings are connected, and the presence of neighboring spaces where dangerous materials are stored.

Pressurization or depressurization is normally achieved by varying the air-flow rates between the air inlets and the returns or exhausts, as shown in Figure 5.19. For example, if a space's thermal load calculation calls for 2,000 CFM (944 LPS) of supply air, sizing the return for only 1,800 CFM (849 LPS) gives a +10% pressurization. But the space's actual, resulting pressure differential is highly dependent on how tight the space is; if very leaky, or with large openings such as entryways without doors, little pressurization, positive or negative, can be achieved. In high-use smoking areas, large openings may be required to accommodate significant traffic flow, but for most other ETS spaces, making the enclosure as tight as possible is probably best. This implies having nonoperable windows, doors

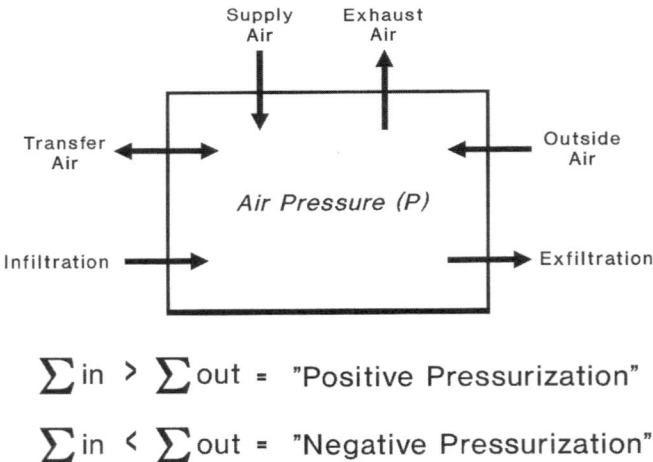

$$\sum \text{in} > \sum \text{out} = \text{"Positive Pressurization"}$$

$$\sum \text{in} < \sum \text{out} = \text{"Negative Pressurization"}$$

Figure 5.19. A certain level of positive or negative pressurization is achieved by setting all the airflows in and out of a space. When mechanical ventilation and exhaust are used, an ETS area should be constructed very tightly so that infiltration and exfiltration are minimized. Exhaust and supply/transfer flows are then adjusted to achieve the needed pressurization, and the space's pressure normally is lower than that of its surroundings.

with automatic closers, air retarders, and making great efforts to seal cracks, and so on during construction. Be aware of somewhat hidden, unintended leakage paths such as duct chases and open plenums.

The "Ventilation and Infiltration" chapter of the *ASHRAE Handbook* (ASHRAE 2001) gives an orifice flow-type method and some coefficients for estimating the air leakage through building envelopes, but the method requires knowing the pressure differential. It is possible to back-calculate the pressure differential if the transfer air rate and leakage paths are well known or predictable. But due to the unknowns associated with actual construction, such estimates are likely to be highly inaccurate. Alevantis et al. (2003) give recommended negative pressurizations from a variety of authorities for various non-ETS hazards, and they range from 0.001 to 0.05 in.w.g. (0.25 to 12 Pa). If a specific pressurization value is required, and with the unknowns of construction, it is probably to your advantage to design your system to be highly adjustable and then to require that the needed pressurization be set in the commissioning phase of the system. Differential pressure gauges are relatively inexpensive, so specifying one for permanent mounting may be a nice feature that you can recommend. For more critical applications, a sensor that triggers an alarm to occupants and maintenance personnel when the desired pressure differential is lost may be in order. Another option is a simple device produced by at least one vendor; it is a clear, sloped, through-wall tube with a brightly colored ball that rolls to various positions depending on the air pressure differential.

If extremely careful control of the pressure differential is in order, such as for very sensitive applications in medical or manufacturing facilities, then the air-flow system should be designed similar to that for clean rooms, hospital isolation units, or laboratories. For example, precision air-flow control dampers on *both* the supply and exhaust air ducts can be specified. With quick-response sensors and actuators, an automatic control system connected to them can then maintain a desired pressure differential. ASHRAE and others have design guides for clean rooms and labs available (e.g., ASHRAE 2002; Whyte 1999). For less critical applications, active control of fan speed can help keep a certain pressure, but due to their significant rotating masses, fans may require longer times to react. Swinging and possibly calibrated, spring-loaded transfer air dampers, doors, or transoms may also help maintain a desired room pressure to counteract when doors and/or windows are opened and closed.

5.7.1. Air Velocities through Open Entryways

When a person leaves a smoking room, he or she will transfer some ETS. The goal, however, is to reduce this contaminated air transfer to as low as reasonably possible. As previously mentioned, self-closing sliding or pocket doors are recommended between ETS and ETS-free areas. But in high-traffic areas, doors may not be practical, and instead one or more open entryways may be desired. In a few rare cases, such as when the whole building is an ETS area and the entryways are to and from the out-doors, loss of ETS out the openings may not be problematic. But when attached to ETS-free areas, care is needed to minimize ETS flows out open entryways. Before addressing the air-flow requirements to contain most of the ETS, some architectural measures should be applied first.

The deeper the entryway, as shown in Figure 5.20, the more time the airflow traveling through it has to organize. Minor flow disturbances, such as the passage of a person, will have more opportunity to settle down and reaccelerate back toward the ETS space. Providing a bell-mouthed entry

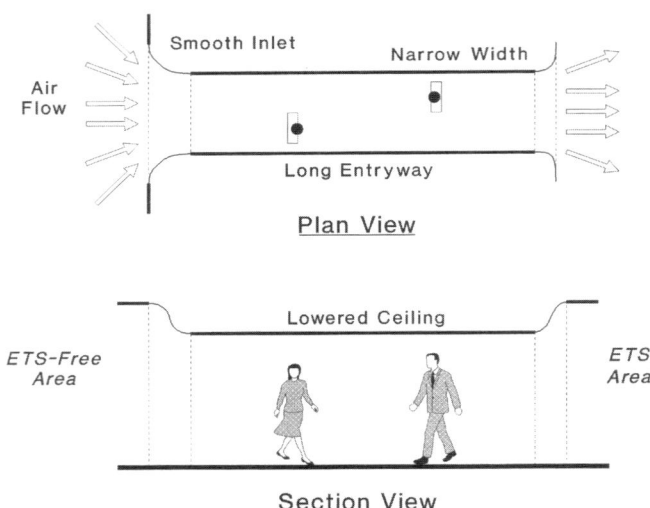

Figure 5.20. Sliding or pocket doors, with automatic closers, should be used on indoor entryways to ETS spaces. However, open entryways are often desired when high foot-traffic is expected. These open entryways can be designed to maximize air velocity in and to reduce ETS migration out. But be sure to investigate fire egress requirements as well.

to the corridor, as much as possible, will reduce separation recirculation bubbles near the ETS-free side of the entryway. Investigate the required fire egress width requirements, but making the entryway as narrow as possible helps, as does lowering the ceiling to increase the air velocity. Providing air dams at the ceiling will help capture buoyant ETS; some exhaust from any coffered ceiling space that is created in the entryway will then help remove this trapped ETS, but if the velocity of the airflow to the room is high enough, the buoyant ETS will eventually be swept back to the room anyway. Be aware that thermally induced air currents, such as created when glass exterior walls are present, can be substantial and increase ETS movement out of a low-velocity entryway.

Once the cross-section of an open entryway has been determined, the next step is to establish the desired average air-flow velocity through it. Some advise that 50 to 100 FPM (0.25 to 0.51 m/s) or more is needed to effectively contain airborne particles inside rooms (ANSI/AIHA 1992); others have suggested that 250 FPM (1.27 m/s) or more is needed to contain superheated smoke (SFPE 1995, ch. 4–12). More research on these velocities for ETS-specific containment is likely needed. From the conservation of mass with steady-state, steady-flow conditions and a constant density assumption,

$$\dot{V}_{air} = \overline{V}_{air} \bullet A_{entryway} \tag{5.2}$$

or, rearranged for the average entryway velocity, as

$$\overline{V}_{air} = \frac{\dot{V}_{air}}{A_{entryway}} \tag{5.3}$$

where the area of the entryway ($A_{entryway}$, ft^2 {m^2}) is evaluated at its smallest vertical cross-sectional plane normal to the direction of the airflow. This smallest area plane is also known as the "throat" of the passageway.

As an example, consider a small smoking break-room that has an open 3 ft × 6 ft 8 in. (0.91 m × 2.03 m) entryway, is otherwise well sealed, and has net exhaust air-flow rate of 2,000 CFM (944 LPS). All of its ventilation air is being provided via transfer air through its open entryway, and exhaust grilles are placed on the ceiling on the opposite side of the room. The area of the entry and the average air-flow velocity through the entryway are then as follows:

$$A_{entryway} = 3 \; ft \bullet 6.67 \; ft = 20 \; ft^2 \qquad (5.4)$$

and

$$\overline{V}_{air} = \frac{2,000 \dfrac{ft^3}{min}}{20 \, ft^2} = 100 \; FPM \qquad (5.5)$$

Note that this is the *average* air velocity across the plane and that it will also vary with time. Significant variations in air velocity at any particular point in the throat will occur due to air temperature differences, pressure fluctuations, passage of users, and many other factors. The supply, exhaust, and other airflows at various conditions must be balanced to ensure the necessary flow through the open entryway, especially when served via variable-capacity systems.

If the pressure differential, but not the flow rate, is known, an orifice model can also be used to estimate the average entryway velocity. But such results should be used carefully as the models' coefficients for large openings are only gross estimates, and, again, the as-built pressure differential may be substantially different from that predicted.

5.7.2. VAV Systems

So far, an inherent assumption, for the most part, has been constant air-flow rates. If, instead, variable air volume systems serve an ETS area and/or its surrounding spaces, the pressure differentials between the spaces will change as the supply air-flow rates vary to meet the thermal loads. Care is needed in the HVAC, exhaust, and control equipment's setup to ensure that the desired minimum flow rates, and needed pressure differentials, are established and maintained.

5.8. ADJUSTABILITY

HVAC design requires making reasonable approximations for many unknowns, such as the occupancy schedules, wattages of equipment, construction quality, and future maintenance. Characterizing ETS in advance is particularly tricky due to occupants' sensitivity to it, as well as changing attitudes and regulations. It is, therefore, wise to design in significant *adjustability* to allow for appropriate commissioning of your ventilation systems and to potentially allow the ventilation air-flow rate to be significantly increased with little or no modifications to the "hard parts" of the systems. In a few cases, occupancy rates or use of an ETS area may prove to be significantly less than that designed for, so potentially the air-flow rate may instead be adjusted lower, with great care, to reduce energy consumption. If an ETS area is later permanently converted to a no-smoking area, this ability to significantly reduce the ventilation air-flow rate can yield dramatic energy savings.

Some techniques for adjustability are providing outside air-flow rate control, supply fan flow-rate control, and even just providing additional rough-ins for future expansion. For a recirculating system, the base %OA was calculated and then specified in the design drawings. By having a high-recirculation system, there is great potential for increasing its 20% or so *OA* upward, via typically interlinked *OA*, *EA*, and *CA* dampers, to add ventilation air. But the air handlers and ductwork must be able to accommodate and treat this additional air. Some, such as many predesigned RTUs, have restricted %OAs, often 50% or so, and may not have the spare thermal capacity needed to condition additional *OA*. Systems that have air-side economizers already have the additional air-flow rate capacity but possibly don't have enough thermal capacity. In humid climates, increasing the *OA* without the needed moisture removal capacity can lead to unacceptable indoor humidity levels, so a psychrometric analysis is needed before any such changes are made to an existing system. ASHRAE's *Humidity Control Design Guide for Commercial and Institutional Buildings* (Harriman et al. 2001) describes this problem and solutions in detail.

Fan control can be used to adjust the flow rates of the *OA* or *EA*. Different adjustment methods include using belt drives so that the sheaves can be replaced to change the fan speed, staging of multiple backflow damper-fitted fans, and various types of adjustable speed drives. Much more information about fan sizing and control can be found in the *ASHRAE Handbook* and in publications from AMCA (e.g., 1990).

If variable capacity is to be provided, consider allowing a degree of direct user control. For example a pushbutton/operator switch, which controls a timed override to increase the ventilation air for 15 minutes or so, could be located conveniently so that a bartender or building manager could intentionally increase ventilation during particularly smoky periods.

In bars, for example, coin-operated overrides could give the customers additional choice too.

Last, and least convenient, but maybe most economical for the owner/operator, is to plan in provisions for *future expansion* of system capacity. This could range from specifying, for example, extra electrical capacity, to increased chilled water pipe sizes, to larger air handlers with empty coil and filter sections for future use. Larger than currently needed filter slots, with smaller frames installed for current use, allow for improving filtration in the future. As of yet unneeded additional wall or floor penetrations for future piping, conduit, and ductwork can be framed out and then recovered to make expanding systems much easier in the future. Be sure that when recovering an opening in a fire-rated surface that the assembly meets or exceeds the needed rating; a note on your as-built drawings explaining the purpose of such "for expansion" provisions would be useful for both code personnel and future modifiers.

5.9. ENERGY CONSERVATION

Except in very mild climates, increased outside air-flow rates indicate that extra conditioning of that air will be required to provide thermal comfort for the building's occupants. The heating, cooling, humidification, and dehumidification processes all require considerable energy use. It is good, therefore, from an energy-cost point of view to make the outside air-flow rate as low as possible. But the opposite is likely true from an IAQ-only perspective, if the *OA* is of good quality. Those lucky few in the United States who are designing buildings for cool climates, or dry but not too hot environments where evaporative cooling is allowed, can have the best of both—high ventilation rates and low cooling energy consumption. For the remainder and vast majority of us, we should examine various methods for making the ventilation air-flow rates as high as possible, also for optimizing energy use.

5.9.1. Good Design Comes First

Good, basic design and optimization are likely the most effective "conservation" measures available. Sizing the ducts larger to reduce pressure losses and selecting efficient fans and motors are examples. Tightly seal an ETS space so that the needed negative pressurization and the air-flow velocity through the entryway can easily be achieved. Also, make sure that the setpoint conditions aren't too cool when in air-conditioning mode (ANSI/ASHRAE 2004c). In heating conditions, operating a smoking area somewhat cool may save some energy and can also possibly increase acceptance of odors and irritants somewhat (Cain et al. 1983). But if the smoking area is a space for working occupants, rather than for visitors or short breaks, then the thermal comfort needed for maximum productivity may prevent lowering the air temperature below normal comfort conditions.

5.9.2. Controls

Next, active engineering measures can be employed to secure more energy savings. For example, *automatic controls* should be used. If the occupancy schedule is very well known, even a simple mechanical or electronic time-clock can be used to *reset* the ventilation rate (ANSI/ASHRAE 2001, addendum *n*). But be sure to consider any after normal business hours occupancy, such as that of security guards, maintenance personnel, and janitorial staff; setting the flow rate to zero may be highly inappropriate. Also be sure to include a vent-out period after occupancy ceases to help flush the space of residual airborne and off-gassed ETS. Including a surge protector and battery-backup system for any electronic control system is advisable to improve its reliability.

An official interpretation of Standard 62 (62-2001-21) advises that simple on/off, residential-like control systems that use air temperature as their control variable are not appropriate for use on ventilation systems covered by the standard. This implies that more complex systems are needed, such as those that measure multiple variables and provide proportional or timed responses. Standard 62.1-2004 does allow for, with great care, periodic on/off controls as long as the ventilation rate is achieved on average; the standard, its official interpretations and addenda, and its new user's guide provide more details. But for ETS areas, such on-off control schemes are inadvisable, because concentrations will grow rapidly during off periods when there is smoking, and pressure differentials may be lost. Significantly decreased acceptance will likely be the result.

At the more complex end of the wide spectrum of control schemes is *demand controlled ventilation* (*DCV*). With DCV, the ventilation rate is adjusted to match the actual or predicted occupancy. A variety of *occupancy sensors* are used with DCV, from light beam/photocell people-counters at doorways, to motion detectors, to contaminant measuring devices. These sensors then feed their near-real-time information to controllers, often microprocessor based, that then send out instructions to devices to, for example, increase or decrease fan speeds or adjust *OA* damper positions.

Selecting sensors for IAQ purposes has been problematic. These sensors are often quite expensive and have required fairly frequent recalibration and/or replacement. Some manufacturers now claim reduced maintenance requirements, but you should evaluate the sensors and all other products carefully. A common placement for contaminant sensors has been in the final exhaust or return duct, so this location implies averaging of the sensed values. For individual control of multiple local exhausts, more sensors will be needed and thus the acquisition and maintenance costs will increase considerably. Placing sensors in-room at breathing height and away from surfaces gives better results for ventilation, but architectural limitations, the potential for unintentional physical damage, and tampering often preclude this placement.

The most common sensor now in use for DCV is for carbon dioxide (CO_2); not only do occupants produce CO_2 but so does the tobacco combustion process. Standard 62's interpretation 62-2001-17 states that CO_2 alone is not an appropriate surrogate for sensing all pollutants—and ETS would be one such large source of various pollutants. Another interpretation, 62-2001-34, clarifies that CO_2 can be used as part of a DCV system, but only if no CO_2 filters are employed; addendum *n* allows adjusting the occupant component. If impregnated activated carbon filters are used as part of the ETS control system, the interpretation implies that CO_2 sensing can't be used (ASHRAE 2003b, ch. 45.9), but others question whether new carbon filters will affect CO_2 levels for very long after installation. Volatile organic compound (VOC) sensors are another type employed; but, again, an interpretation (62-2001-32) says that VOC sensors alone are not acceptable. ETS sensors have been marketed, but there are various types and qualities. Evaluate each to determine, for example, if they are just CO_2 or general VOC sensors or if they are specialized to one or more of ETS's constituents. There are various compounds associated with ETS; some ETS sensors can reliably detect some of these compounds, but there is significant debate over whether any one particular compound and, therefore, any particular sensor is appropriate.

With any shut-down DCV system, a significant concern arises during initial occupancy, such as in the early morning of a typical workday. The evening before, the DCV system kept the ventilation rate high until the tracer was reduced to the lower end of the throttling range, and then the admission of outside or cleaned air was ceased. Overnight, off-gassing of building materials and ad- and absorbed ETS has continued, however. Upon first entry, and depending on the production rate, mixing, and interior air volume, it may take hours before the concentration of the particular tracer gas becomes high enough to cause the DCV controller to start admitting ventilation air. During this period, occupants can be exposed to locally excessive levels of various contaminants and odors. To reduce or avoid this problem, DCV systems now often include a base ventilation rate, high-volume initial flush-out periods, and lower turn-on setpoints. Also be sure that DCV-controlled ETS exhaust systems don't slow or shut down before they have had sufficient time to remove the residual, low concentration ETS after occupancy ceases. If the negative pressurization is lost via too early DCV shut-down, significant ETS will migrate to neighboring spaces. A local exhaust system can be operated separately from the building's central DCV system, however, to continue venting contaminants.

5.9.3. Heat Recovery

When high flow rates of different temperature air are being admitted to and expelled from a building or industrial processes, heat recovery is often a very effective but somewhat expensive to implement energy conservation measure. For buildings, the expelled air is normally at or very near the desired indoor air conditions—for example, 75°F (23.9°C), 50 percent RH. The incoming outside air is at or near the ambient conditions, and in many parts of the United States can range from very cold (for example, 0°F [–17.7°C] at design conditions) to very warm (98°F [36.7°C], for example). If there are a substantial number of operating hours where the intake-to-exhaust temperature difference is high, maybe 20°F (11°C) or more, then the economics for providing heat recovery may be favorable. In cold weather, heat from the exhaust air can prewarm the outside air somewhat, and in the summer transferring heat from the intake to the exhaust can precool the outside air.

Heat exchangers (HXs) are needed to perform this and other types of heat recovery, and there are many available. When possible, they should be installed in *counterflow* rather than *parallel flow* to achieve their highest heat exchanger effectivenesses, but many are restricted to *cross-flow* by their nature. *Direct-contact heat exchangers* allow a heat source or sink to be directly in contact with the heat transfer medium—for example, water. A wet cooling tower is a direct-contact heat (and mass) exchanger as the water is cooled directly in the tower's outside air stream via convection and evaporation. An *indirect heat exchanger*, such as a heating or cooling coil, provides a physical separation that keeps the two flows distinct, and typically cleaner, but at a somewhat reduced heat exchanger effectiveness.

With all applications, but especially for ETS, it is important that HXs have ready provisions for inspection and, when needed, for periodic cleaning. Both airflows, the outside air as well as the exhausted air entering an air-to-air HX, need *filtering* to help keep the heat transfer surfaces clean. The following text describes the most common heat recovery heat exchangers used for ventilation purposes, and some comments are given on their potential performance when ETS is present. While the emphasis is on reducing energy consumption, heat recovery heat exchangers can also be employed to dramatically increase the outside air-flow rate while keeping about the same overall energy consumption; a combination of increased outside air-flow rate and somewhat reduced energy consumption is also possible.

5.9.3.1. Plate-type Heat Exchangers

A *plate-type heat exchanger*, shown in Figure 5.21, is made of many layers of a thin-walled heat transfer material, such as aluminum, and is a popular option for *OA* heat recovery. This type of HX is classified as an air-to-air HX, but it requires that the two air streams be near enough to each other so that air pressure losses and duct heat losses/gains are low. Most plate-type HXs have solid heat transfer materials, so only sensible ("dry") heat is transferred between the air streams. With good construction and a side-to-side pressure balance, little to no moisture or contaminants move from one air stream to the other; this is critical where codes, or the designer, require that no significant ETS be present in the ventilation air stream.

The sensible heat-only, plate-type heat exchangers often have heat exchanger effectivenesses of about 50%, which is quite good. Some versions have a porous heat transfer membrane, which allows mass transfer; moisture (latent heat) and presumably some contaminants can cross. These versions often claim effectivenesses up to 70% or so. Due to the potential for movement of contaminants from the exhaust to the incoming

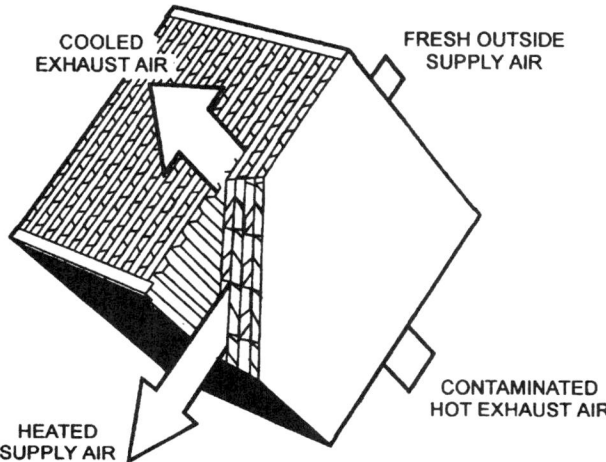

Figure 5.21. A fixed-plate or plate-type air-to-air heat exchanger (HX). These HXs are commonly used in residential-type energy recovery devices, as well as for some commercial applications (ASHRAE 2004, ch. 44.9).

air, and for increased fouling of the exhaust side of the HX, they may not be as attractive for ETS applications as sensible heat-only versions of these units.

5.9.3.2. Rotary Heat Exchangers

Rotary heat exchangers, or "heat wheels," are interesting hybrid heat transfer devices, and one is shown in plan and elevation in Figure 5.22. A large wheel of porous heat transfer medium rotates slowly from one air stream to the other and moves heat in the process. Sensible heat-only versions exist, but often the medium is treated with a hygroscopic material, which absorbs and then re-emits moisture from one side into the other. With the latent heat transfer via these desiccant coatings, in addition to the sensible heat transfer, peak heat exchanger effectivenesses of 60% to 70%, or even 80%, are common. This type of HX typically has a very high capacity. They are often employed in industrial applications, such as for preheating combustion air with heat recovered from hot flue gases. Similar to the plate-type heat exchangers, they require that the two air-flows be relatively near each other, so they may not be practical in appli-

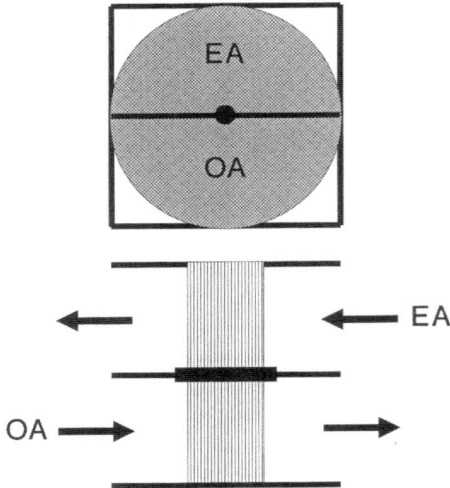

Heat Wheel Heat Exchanger

Figure 5.22. A heat wheel heat exchanger uses a porous, slowly rotating media to move sensible and possibly latent heat between the air streams.

cations where, for example, an MAU is on one side of a building, and the relief fan is on the other. As the HX wheel is motorized, some electricity is required for operation, but capacity control is possible by adjusting the speed of rotation.

The large heat wheel media require periodic replacement, so positioning the unit for ease of future O&M is important. And, as with the Wankel engine, these rotary HXs have the same challenge in sealing the rotors' sides to reduce leakage. While in the past 3% or so cross-contamination was common, it is now possible to reduce this to 0.5% or so. Often a vented gap in the HX housing between the hot and cold air streams is used to further reduce this air and contaminant transfer, or "carry-over," but at a cost of slightly reduced effectivenesses. Again, if absolutely no ETS in the outside air stream is a goal, and is a very restrictive and somewhat unrealistic requirement, then due to this leakage between the sides, a heat wheel may not be the best choice. But if a little leakage is acceptable, and often is, with this heat recovery you can increase the outside air-flow rate a bit to increase ETS dilution and still not incur a net energy penalty. Filtering both air streams before they enter a rotary heat exchanger greatly extends the wheel's life, and, for ETS applications, using a HEPA filter on the exhaust air stream is desirable.

5.9.3.3. Run-around Heat Exchangers

Run-around heat exchangers, as depicted in Figure 5.23, are two or more liquid-to-air coils that are connected via piping, a pump, and a liquid heat transfer medium. As this equipment is often outdoors, and the outside air coil is exposed to cold air at times, the heat transfer fluid is usually a water-glycol-inhibitor solution. In frost-free climates, distilled water might be used. Due to the increased quantities of heat transfer materials between the two air streams, and the needed pump energy, heat exchanger effectiveness claims tend to peak at around 50%. Only sensible heat can be transferred, but by using a detailed energy model, Dhital

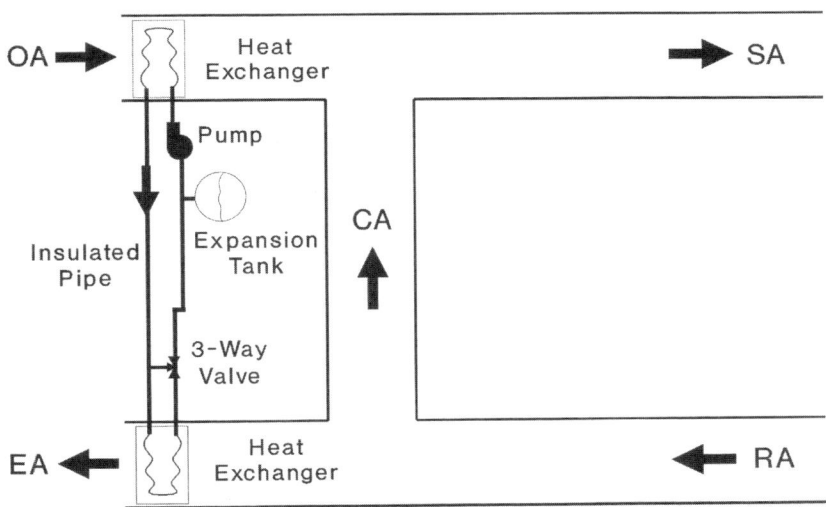

Run-Around Heat Exchanger

Figure 5.23. A run-around heat exchanger uses two water-to-air HXs, a pump, piping, and a heat transfer fluid such as water/glycol to move sensible heat. A three-way valve, installed in one of various places, is often incorporated to provide some freeze protection for either or both coils, as well as capacity control when using a fixed-speed pump.

et al. (1995) found that a large office building situated in four different U.S. cities could reduce annual electrical energy use up to 3.5% and natural gas consumption by up to 41%. These types of HX systems are not normally sold as units but instead are specified piece by piece by the design engineer.

Run-around heat exchangers offer several advantages: the air streams can be separated, often by 100 to 300 ft (30 to 91 m) before the pumping energy requirements become excessive; contaminants can't be transferred; and the pump can be controlled to vary capacity. A three-way valve is usually installed, as shown in Figure 5.23, to keep liquid flowing through the outside air coil for enhanced freeze protection and to allow the use of a constant speed pump, if desired. An article about sizing and operating run-around heat exchangers is available in the *ASHRAE Journal* (Besant and Johnson 1995).

5.9.3.4. Spray-type Heat Exchangers

"Chemical" or *spray-type heat exchangers* are direct-contact HXs, as shown in Figure 5.24. They typically use a sorbant-type liquid such as a lithium-bromide solution to move sensible and/or latent heat from one air stream to another. As they are direct-contact, typically use unpleasant chemicals, and often require considerable maintenance, they may not be attractive for all but the largest ETS heat recovery applications. However, these types of HXs are often employed as part of fume scrubbers, so they may find use for large ETS-laden exhaust air flows where regulations require treatment before discharge.

5.9.3.5. Heat Pipe Heat Exchangers

Heat pipe heat exchangers are elegantly simple devices and can be highly effective at transferring heat. Some limitations are that the airflows must be adjacent, only sensible heat is transferred, and while each approximately 1 in. (2.54 cm) diameter pipe is not too costly, many are needed, so the total cost may be prohibitive. As shown in Figure 5.25, heat pipes are tubes, often copper, fitted with a wick and then sealed, evacuated, and charged with a refrigerant. When installed, the refrigerant then evaporates on the warm end and condenses on the cool end. The wick helps move the condensed refrigerant back to the warm end of the pipe. For optimal per-

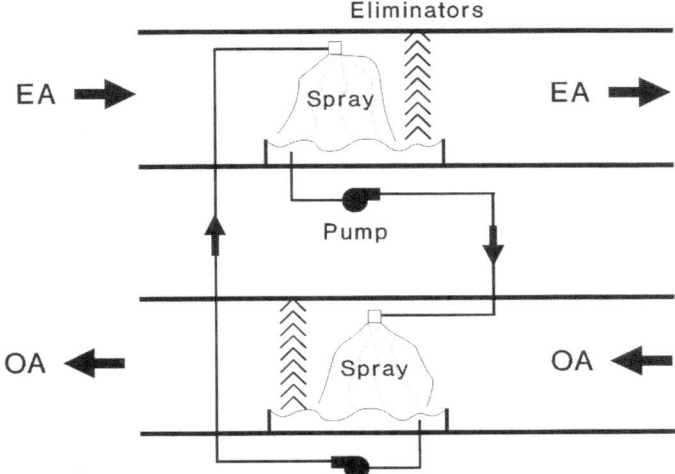

Spray-Type Heat Exchanger

Figure 5.24. Spray-type heat exchangers are direct-contact and often of very high effectivenesses, but they typically employ chemicals that are not well suited for most occupancies and some maintenance staffs' expertise.

formance, the pipes need to be slightly tilted, and this tilt needs to change at least twice yearly in most U.S. climates. But with an appropriately sized heat pipe heat recovery unit, often utilizing extended surfaces ("fins") to increase its capacity, a high proportion of the sensible heat can be reclaimed.

Placing any heat exchangers, and likely panel filters to protect them, into the air streams introduces more air-side pressure losses, often about 0.3 to 1.0 in.w.g. (75 to 249 Pa). As such, fan power will likely need to be increased but can be minimized with optimal design. Sizing any of the preceding types of heat exchangers requires considerable knowledge of heat transfer phenomena and an iterative selection process. As such, new entrants to the field should obtain help from senior engineers who have experience in HX selection. The *ASHRAE Handbook* and various heat transfer textbooks contain much of the basic theory and other information needed, but practical design software and manufacturers' literature will likely be required as well. The difficulty with specifying heat recovery HXs, and the gradual loss of senior designers' experience, is likely the

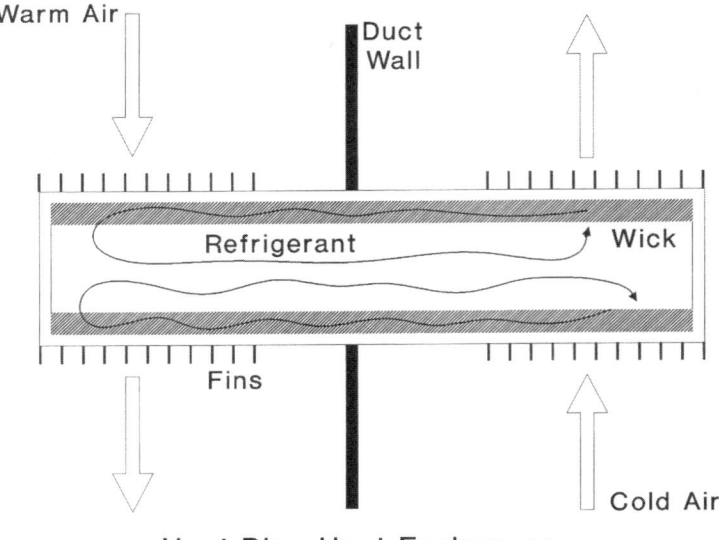

Heat Pipe Heat Exchanger

Figure 5.25. Heat pipes are elegant, sealed devices that move sensible heat via evaporation and condensation of a refrigerant.

reason why these devices, with vast energy-saving potential, are not utilized nearly as often as they should be. As such, there may be significant opportunity for you to become your office's in-house expert on heat recovery HXs and similar devices.

5.9.4. Energy Guidelines

The ventilation systems used for ETS applications need to follow mandatory energy conservation requirements, such as the various versions of ASHRAE Standard 90, as adopted in the particular location. The heat recovery measures just described can help offset the increased energy use associated with the enhanced airflows needed for handling ETS. The elective Leadership in Energy and Environmental Design (LEED) program of the U.S. Green Buildings Council (www.usgbc.org) includes credits for these advanced energy systems, as well as requirements for buildings if ETS is to be present.

5.10. COMMISSIONING, OPERATION, AND MAINTENANCE

Commissioning and *operation and maintenance (O&M)* issues are often not given much consideration beyond the specifications stage of the design process, but they are crucial to the long-term success of any HVAC system. After a system is finished, the mechanical subcontractor typically starts up the system and does some initial adjustments. But "commissioning" implies that much more is done in the way of planning, tests, and checks to make sure that the system will operate as intended under a variety of conditions. Stages of this commissioning are done before, during, and after completion of a system (ASHRAE 2003, ch. 42). A test, adjust, and balance (TAB) technician usually performs the commissioning and relies heavily on the designer's drawings and specifications. For ETS applications, ensuring that ventilation systems perform to specs is critical. Should the as-built system underperform, corrective actions must be taken. The causes for the underperformance and the postconstruction measures should be well documented.

In the design phase it will be a long-term benefit of the system, occupants, and owners/operators if you consider future maintenance needs. For example, installing larger filters usually extends the period between necessary changes and can reduce disruptions. Mechanical equipment should be placed as accessible as possible for inspection, preventive maintenance, and necessary repairs. For example, fan-coil units should be positioned so that the controls, piping, filter access, and other potential maintenance items are not sandwiched up against walls but rather are on the room side of the equipment. Requiring a tall, portable ladder to replace filters will, over time, likely decrease the frequency of their replacement; where possible, installing a catwalk or repositioning the filter access will encourage regular inspections and needed maintenance. If noise is otherwise controlled, specifying external (wrap) duct insulation rather than lining will allow easier duct cleaning. Providing access hatches in critical locations for such cleanings will assist in future maintenance. For a readily cleanable characteristic, metal ductwork is advisable, but some linings on rigid ductboard and other types of duct materials may also be rugged and repeatedly cleanable to an acceptable degree. Provide access doors in the air handlers too so that all portions may be inspected and cleaned when needed. As the coils may need more frequent cleaning when exposed to ETS in recirculating systems, consider specifying thicker finned rather than thinner versions. Require the contractor to replace the air filters just before turnover to the owner. Also provide at least one spare set of new filters and their acquisition information to help ensure that correct replacements will be obtained and installed in the future by the owner/operator.

In the ETS areas themselves, architectural measures should be employed to assist in O&M. If possible, all exposed surfaces should be hard, fire resistant, and able to withstand repeated cleanings. Carpet and "fuzzy" wall coverings should likely give way to smooth floor coverings and painted walls. Suspended ceiling tiles are often made of highly ad- and/or absorptive base materials, such as recycled newspaper or fiber-glass, so installing a painted "finished" ceiling of gypsum board or metal, for example, is preferable. Glossy oil-based paints tend to be more clean-able and durable than latex. When selecting furnishings and finishes, care is needed to reduce the chance or severity of fires as some tobacco prod-ucts, lighters, and matches are common ignition sources (SFPE 1995). Be generous in specifying appropriate portable fire extinguishers in and around smoking areas. Automatic sprinklers, even if not required by code or installed elsewhere in the building, should be considered for smoking areas. Floor drains, with adequate slopes, cover grates, and trap primers would be helpful too for floor moppings and for drainage should sprin-klers activate; these may or may not be required by code.

5.10.1. Standardize O&M Procedures

To increase the reliability of any HVAC system, trained maintenance staff are needed, inspection and maintenance procedures need to be defined, and the preventive work needs to be completed regularly. If the owner's staff is unfamiliar with a certain advanced aspect of your design, such as the use of HEPA filters or DCV, then they should obtain the necessary training, or an experienced technician should be hired or otherwise retained. The system needs documentation too; require that the contrac-tors provide the owner with all the original equipment manuals organized neatly in a binder. If not done by a contractor, the owner needs to establish and document maintenance procedures, schedules, lists of common replacement parts, and have suppliers' contact information ready. Having maintenance staff keep records (e.g., checklists) of what and when main-tenance was done is important, especially for ETS applications. Owners are advised—for example, through the designer's specifications—to acquire the needed supplies in advance, such as having spare filter sets, belts, and small motors on hand, because special situations such as equipment failure or higher than expected use may require early or unexpected maintenance. As vendors sometimes come and go, keeping spares on hand will also help keep the facility in operation during supply-chain difficulties. Much more information on good O&M practices can be found in Chapter 38 of the *Applications* volume (ASHRAE 2003) and elsewhere.

5.10.2. Signage

Ordinances or future standards may require signage outside each public entrance to ETS areas, and these signs likely must at least state "This Area May Contain Environmental Tobacco Smoke" or similar, as shown in Figure 5.26. At least 1 in. (2.54 cm) tall lettering is recommended, or as otherwise required by applicable code. Other methods of providing this information may be allowed—for example, pictograms or recorded announcements may be appropriate alternatives. Americans with Disabilities Act (ADA) and other requirements may also apply.

THIS AREA

MAY CONTAIN

ENVIRONMENTAL

TOBACCO

SMOKE

Figure 5.26. It has been proposed that Standard 62.1 require a warning sign to be placed outside an ETS area.

5.10.3. Reuse of a Previous ETS Area

As ETS requires some time to be cleared from an air volume, it is not possible to immediately convert an ETS area to an ETS-free space. For example, a recently used designated smoking room in a restaurant cannot immediately be used as a nonsmoking space—ventilation, cleaning, and time are needed for the airborne ETS to be evacuated and the ad- and absorbed product to off-gas and be removed to an appropriate level. More research is needed on this reclassification of spaces, but the more densely and the longer that smoking occurs in a space, the lower the ventilation rate, and the more porous the materials, the greater the vent-out period will likely need to be. Direct observation is probably needed to determine this period for each real space.

5.11. ECONOMIC ANALYSES

Virtually all private-sector engineering projects require at least minimal consideration of their economic viabilities, and ETS applications are no exception. Building owners and architects prefer having options, so evaluating and then presenting a reasonable number of alternatives is advised. One alternative is a complete indoor smoking ban, and this may prove to be the most economical in many cases. Simple payback (SPB) and life-cycle cost (LCC) economic analytical methods are popular for evaluating alternatives and are usually employed for nongovernment jobs, but many other methods exist. For public projects, the cost/benefit ratio (C/R) technique is commonly used. The *ASHRAE Handbook* (ASHRAE 2003a, ch. 36) has much more information on evaluating owning and operating costs.

5.12. ENGINEERING ETHICS

This section and all other portions of this book are not intended as a substitute for legal or insurance advice. Owners, designers, and installers should consult their legal, risk management, and insurance advisors for specific information pertaining to their practice issues and liability.

As good HVAC designers, who try to provide our clients and building occupants with acceptable indoor environments, ETS and other pollutants challenge us. A building owner wants to meet the needs of employees, tenants, and/or customers. Allowing indoor smoking in all or part of a particular building may be a requirement for a project and is a decision made exclusively by the owner unless laws to the contrary are in force. Yet Professional Engineers are also obligated to help protect the public's health and safety (NSPE 2003). Having ETS present is arguably at odds with this second goal (Glantz and Schick 2004). Due to human behavior and other factors, people will be exposed to some ETS at times, even in designated ETS-free spaces and buildings, no matter which design steps are taken. We are, however, capable of designing systems that provide significant odor and irritation control for most hours of a typical year.

In the earliest stages of your design process, you will need to obtain from the owner and the architect a list of the desired ETS areas. Inform them that if indoor smoking is to be allowed, you can only seek to address airborne ETS odors and irritants and not health; your designs cannot prevent any occupants, whether visitors, customers, or employees, from some level of exposure to ETS. Investigate with your legal counsel and insurers whether obtaining a contractual release and an indemnity clause is in your best interest (ASHRAE 2003c). Determine whether your professional insurance covers your ETS-related work. An introduction to engineering law, liability, and ethics is available in Morton (1983) and other related publications. Brief articles also often appear in the *ASHRAE Journal* and other trade periodicals.

Your efforts as an HVAC designer to improve IAQ in buildings should be of great benefit to all involved. Each project will have different requirements and will likely challenge you. In the next chapter, discussion and examples for various applications are presented to assist you in making careful ventilation decisions.

6

APPLICATIONS

This chapter discusses specific occupancies in which secondhand smoke has been a common concern, such as offices, restaurants, hotels, casinos, and prisons. Also presented are many example ventilation rate calculations, with at least one given per application. Contaminant removal effectivenesses greater than 1.0, ETS air cleaning, and heat recovery are illustrated as well. At the end of the chapter are two tables (6.1a and b), in I-P and SI units, that summarize many examples' results on a per person basis—care is needed in using these results because there are many assumptions included in each value. If reading the entire chapter, many of the provided ETS Dilution Method (EDM) examples will seem redundant, but they are all presented, since future use of the chapter will likely be for specific applications only (e.g., designing a bar); having an example present in each section is to encourage users to adjust the input factors and to recalculate the flow rates, rather than just using the chapter's Tables 6.1a and b values directly. Research on ETS in low-rise residences is underway (e.g., Nazaroff and Singer 2002), so hopefully this book can be expanded soon, or a similar separate design document produced, to address ETS in low-rise single- and multifamily homes.

6.1. OFFICES

Office buildings vary in size from, for example, a small one-person insurance agent's place of business, to massive skyscrapers full of workers. While the smallest buildings are often occupied by just one business, mid-to-large size buildings normally have space for multiple companies. Many of these businesses lease their office areas, and they frequently expand into more space in the same building or move into a large facility as their product or service sales and/or workforces increase. Each company in a building is likely to have its own smoking policy—each must be designed to, unless a stricter local or state code is in force, or the building owner/operator imposes a stricter policy through the leases.

If only one company is to occupy a new building, and preferably owns it, designing the HVAC systems for long-term use has less uncertainty than if the building's occupants are transient and/or unknown. Many office buildings are built on a speculative basis, so the future tenants' needs are uncertain; designing the building's central systems first and then the 'tenant finish-outs" later is therefore challenging. If possible, including some flexible provisions for floor-by-floor exhaust—for example, roughed-in duct chases or places in the envelope for exhaust louvers, will help in designing systems for the as of yet unknown occupants. In rental spaces with existing HVAC systems, new occupants often reuse the systems without major modifications; these systems should, however, be evaluated by HVAC engineers and others, and, if need be, the systems should be redesigned.

The type and quantity of building users will vary significantly from company to company. Many service and catalog firms these days do their work via the mail, phone, or the Internet, so customers rarely enter their buildings; in such cases designing for the employees, only, seem appropriate. But many businesses have various degrees of customer traffic flow through their offices, and these visitors should be given appropriate consideration. It's likely that a small portion of the company's floor space is used by customers, so only these parts of the facilities may need the extra ventilation needed to improve acceptability for visitors. For example, a tall bank building will have many visitors in its street-level lobby; a few on the floors where private meetings occur; and maybe none on its data, check, cash, and securities processing floors.

Variable air volume systems are commonly used in office buildings in the United States. An exception is that smaller buildings may use constant air volume AHUs or RTUs, especially if the buildings have only a few thermal zones each. Larger office buildings will likely have dozens, if not hundreds, of zones, and thus VAV becomes very attractive for meeting the thermal needs of these differing spaces. In designing for ETS, using transfer air and local exhaust seems appropriate for small parts of these buildings, but for large areas, general dilution may be appropriate. Be sure not to recirculate or transfer air from ETS areas to non-ETS spaces; consistently maintaining desired pressure differences and air-flow rates with single-duct VAV systems will be very challenging.

When the occupants, their desires, and policies, and the ETS and ETS-free areas are defined, it is then possible to optimize your HVAC designs. Slightly different approaches are needed throughout a building to address different densities of expected smoking. A general office area will likely have fewer cigarettes smoked per hour per unit floor area than a smoking break-room, for example, so the latter will need much more ventilation air per occupant. Open office space, enclosed single-person offices, and conference rooms are considered in this section, but smoking-allowed break-rooms are presented in a later section of the chapter as they can be used in other types of buildings as well.

6.1.1. Open Floor Plans

While the perimeter of an office building is usually furnished with enclosed, single-person offices, the center and usually windowless portions are often somewhat open work areas. Many times these open areas are furnished with short dividers to create semiprivate work areas, but a common HVAC system normally serves these areas. If smoking is to be allowed in this open area, likely only a small proportion of the occupants will be smokers. In a large workgroup, the portion should be similar to that of the local population who are smokers. Exceptions do occur, of course, due to company activities, demographics, and so on. A tobacco company, for example, might have a much larger percentage of smokers, whereas a health-care organization might have few or no smokers on staff and likely won't allow indoor smoking.

In the following sample calculations, an open office area needs general dilution ventilation to control ETS and other contaminants. Two cases are considered for each: having all long-term adapted occupants, or only short-term unadapted visitors. It is possible that a combination of the two extremes is desired for a particular space; further calculations are needed and likely will result in a per person ventilation rate between the two cases.

EXAMPLE 6.1

What is the ventilation air-flow rate ($\dot{V}_{tot}$) needed for a particular 10,000 ft^2 smoking-allowed general office space? Perfect mixing is assumed, all the floor area is occupiable, and cooling is the dominant mode.

Solution:

The first step is to find the base ventilation air-flow rate from Standard 62.1. For this open-plan office space a value of $V_{vrp}/p = 17$ CFM/p is found from the standard's Table 6.1 "office space" entry. The table's estimate for the occupancy in the space is 5 people per 1,000 ft^2, thus there are about 50 occupants of this space during design conditions. Not having more specific occupancy data available, the Ventilation Rate Procedure's base flow rate ($\dot{V}_{vrp}$) is estimated as

$$\dot{V}_{vrp} = 10,000 \ ft^2 \bullet \frac{5 \ p}{1000 \ ft^2} \bullet 17 \frac{CFM}{p} = 850 \ CFM$$

This 850 CFM is the base, non-ETS ventilation air-flow rate required to ventilate the space for all other contaminants *and may require further adjustments due to system performance*, for example, as described in Standard 62.1. The flow rate needed for diluting the ETS must be found and added to this base value.

The next step toward this goal is to estimate the percent of occupants who are smokers. A value of 20% smokers is selected from Table 4.2 of this book that gives a range of 0.2 to 0.25 for X_{sm} for "all other occupancies." Because this is a work area, the occupants in this office-space example will be assumed to be long term and thus adapted. Table 4.1a then provides the volumes of air needed to dilute the smoke as $V_{cig,ns} \approx$ 3,900 ft^3/cig and $V_{cig,sm} \approx 1,100$ ft^3/cig. Using equation 4.19, the adjusted volume of dilution air per cigarette is then

$$V_{cig} = 0.2 \bullet 1100 + (1 - 0.2) \bullet 3,900 = 3,340 \frac{ft^3}{cig}$$

Table 4.2 of this book provides guidance on the smoking rate, and a value of $R_{sm} = 0.6$ cig/p_{sm}·h is suggested for this occupancy. As the total occupancy (P_{tot}) is 50 p, the design smoking density is, therefore,

$$\dot{D}_{cig} = \frac{0.2 \dfrac{P_{sm}}{p} \cdot 50\, p \cdot 0.6 \dfrac{cig}{P_{sm} \cdot h}}{10,000\, ft^2} = 0.0006 \frac{cig}{h \cdot ft^2}$$

As the space is well mixed, the contaminant removal effectiveness, E_{cr}, is assumed to be 1.0. From equation 4.17a, the extra ventilation air needed to dilute the ETS to acceptable odor and irritant levels is then

$$\dot{V}_{ets} = \frac{0.0006 \dfrac{cig}{h \cdot ft^2} \cdot 3340 \dfrac{ft^3}{cig} \cdot 10,000\ ft^2}{60 \dfrac{min}{h} \cdot 1.0} \approx 334\ \ CFM$$

The total ventilation air-flow rate required for this space, from equation 4.16, is then

$$\dot{V}_{tot} = 850\, CFM + 334\, CFM = 1,184\, CFM$$

or about 139% of the base requirement for a similar nonsmoking space. For this particular space, the ventilation air-flow rate per person is

$$\frac{\dot{V}_{tot}}{P} = \frac{1,184\, CFM}{50\, p} \approx 24 \frac{CFM}{p}$$

and is thus 7 CFM/p more than the required 17 CFM/p base ventilation rate.

EXAMPLE 6.2

What if Example 6.1's occupants have instead just entered the space?

Solution:

In this case, the occupants would be unadapted. The procedure is the same, except the unadapted ventilation air volume for dilution is required. The base ventilation air-flow rate is still 850 CFM, and the same percentage of smokers (20%) is assumed for this example. But because the occupants are unadapted, Table 4.1a then provides the volumes of air needed to dilute the smoke as $V_{cig,ns} \approx 5,600$ ft^3/cig and $V_{cig,sm} \approx 1,400$ ft^3/cig. Again using equation 4.19, the adjusted volume of dilution air per cigarette is then

$$V_{cig} = 0.2 \cdot 1,400 + (1 - 0.2) \cdot 5,600 = 4,760 \frac{ft^3}{cig}$$

Using the same smoking rate of $\dot{R}_{sm} = 0.6$ cig/p$_{sm}$·h and total occupancy (P_{tot}) of 50 p, the design smoking density remains 0.0006 cig/h·ft^2. The contaminant removal effectiveness, E_{cr}, is assumed to remain as 1.0. So from equation 4.17a, the extra ventilation air needed to dilute the ETS to acceptable odor and irritant levels for the unadapted occupants is then

$$\dot{V}_{ets} = \frac{0.0006 \dfrac{cig}{h \cdot ft^2} \cdot 4,760 \dfrac{ft^3}{cig} \cdot 10,000 \ ft^2}{60 \dfrac{min}{h} \cdot 1.0} \approx 476 \ CFM$$

The total ventilation air-flow rate required for this space, from equation 4.16, now is

$$\dot{V}_{tot} = 850 \ CFM + 476 \ CFM = 1,326 \ CFM$$

or about 156% of the base requirement for a similar nonsmoking space. For this space with unadapted occupants, the ventilation air-flow rate is

$$\frac{\dot{V}_{tot}}{P} = \frac{1,326 \ CFM}{50 \ p} \approx 27 \frac{CFM}{p}$$

and is thus about 10 CFM/p more than the required 17 CFM/p base ventilation rate. At the end of this chapter, Tables 6.1a and b summarize the results of this and all the example calculations in this chapter.

6.1.1.1. Adapted versus Unadapted?

Examples 6.1 and 6.2 demonstrate what may be a design dilemma: do you design for first entry of workers or wait for when they are adapted? Good engineering judgment, with base information from the owner, occupants, and architect, should be used. For example, if the workspace has been unoccupied overnight, and was properly vented of ETS, then it will take some time for the new ETS to reach the steady-state concentrations implied by the ETS Dilution Method. As such, hopefully the acceptance will be high as the ETS level increases, but people are becoming adapted. But if the space was already occupied, perhaps by a previous work shift, then designing a fixed flow rate for unadapted occupants may instead be appropriate. It is also possible to increase and then decrease the ventilation rate with time to make an adjustment for when occupants become adapted. But the rate should also be increased well before the next contiguous group enters the space as they will be unadapted and time is needed beforehand to reduce the concentration of ETS to their acceptable level.

EXAMPLE 6.3

Smoking is to be allowed throughout a small, freestanding office building. The general dilution ventilation requirement, from ANSI/ASHRAE Standard 62.1-2004 for the building and occupants, is 1,200 CFM_{oa}. From the ETS Dilution Method, for adapted occupants, another 710 CFM_{oa} are needed for the ETS, so a total of 1,910 CFM_{oa} is to be admitted through the central constant air volume air handler, assuming no ETS air cleaning. The supply air-flow rate is 9,500 CFM, and the HVAC designer plans for "+5%" building pressurization. The selected AHU can, however, accommodate six-inch deep cartridge filters, and the designer locates filters for it that are 90% effective in removing ETS. If such filters are used in the AHU, what flow rate of outside air needs to be admitted?

To meet the VRP general ventilation air requirements of Standard 62.1-2004, at least 1,200 CFM_{oa} will need to be admitted. Next, the credit for using treated recirculated air is investigated by the designer. As the building is to be pressurized "+5%," the return air-flow rate is 5% less than the supply, or $9,500 \times (1\text{-}0.05) = 9,025$ CFM. With the 1,200 CFM_{oa} being admitted for general ventilation purposes, $9,500 - 1,200 = 8,300$ CFM of air is to be recirculated; $9,025 - 8,300 = 725$ CFM is therefore to be exhausted, and $9,500 - 9,025 = 475$ CFM should exfiltrate. With the 90% ETS filtration system in the recirculated or mixed air stream, from Section 4.3.7,

$$\dot{V}_{ets,f} = E_f \cdot \dot{V}_{ca} = 0.9 \cdot 8,300 \, CFM = 7,470 \, CFM \gg 710 \, CFM$$

So if all the recirculated air is filtered for ETS, no additional outside air is indicated, unless needed by other requirements. But the ETS filter will be large, expensive, have a high pressure loss across it, and may require frequent replacement. The filter and fan energy costs will be high.

If instead a parallel fan-powered ETS filtration system is employed that treats only 10% of the recirculated air,

$$\dot{V}_{ets,f} = 0.9 \cdot 8,300 \, CFM \cdot 0.1 = 747 \, CFM \approx 710 \, CFM$$

Thus, this just meets the ETS removal needs as estimated by the EDM, the filter effectiveness, and the designer's assumptions. The fan and filter costs should be much lower than if all the recirculated air were to be treated, but much less $\dot{V}_{ets,f}$ is provided.

Additional information on air cleaner effectiveness for ETS and other applications can be found in Bohanon and Nelson (1999), Nelson et al. (1999), and similar publications.

6.1.2. Single-Person Offices

Individual, enclosed office spaces are often 100 to 130 ft^2 (9.3 to 12.1 m^2) in floor area in the United States, and many are located on the perimeter of buildings to have windows for visual relief and possibly for natural ventilation. When offices are placed on the exterior of a building, local, individually controlled exhausts for ETS may be installed easily if envelope penetrations are possible.

You may be designing for the common situation where a single, valued employee, or company owner, smokes, but there is a general smoking ban in the building. To accommodate this highly valued person, a separate ventilation system can be employed to allow smoking in his or her office and to minimize exposure to others, as shown in Figure 6.1. A local exhaust system that runs the room at a negative pressure may be appropriate. Any existing return air provisions to the central air handler must be blocked and extremely well sealed to prevent recirculation of ETS. The supply air, as well as some transfer air, should be used to make up for the exhausted air. Often, several offices are grouped together in one thermal zone, so in a retrofit situation, you may or may not have full control of the supply air; adding another VAV subsystem may be necessary.

You may instead be designing the HVAC system for an office building that is designated in full as an ETS area. In this case, dilution ventilation of

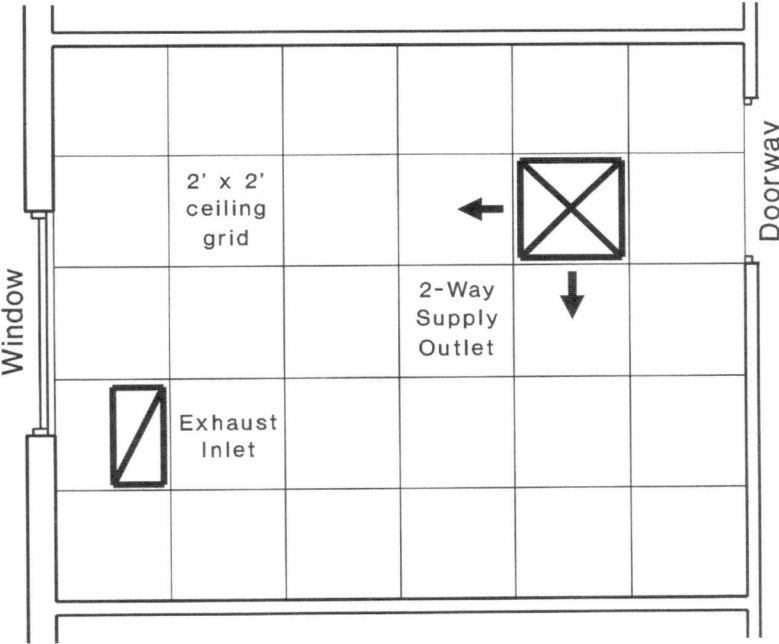

Figure 6.1. Ceiling plan of a single-person office; an existing space has been adjusted for use as an ETS area. Note that a new two-way diffuser, replacing a centered four-way blow terminal, is used to aim supply air into the room and not out the door. The exhaust is placed at the far side of the room to draw ETS away from the occupant, as well as visitors and the doorway.

the individual offices may be more appropriate, rather than direct exhaust. In some cities or companies, smoking is allowed in individual offices only, so you may have single-person offices designated as smoking areas while the immediately adjacent halls, open areas, and support spaces may be ETS-free spaces. Example 6.4 finds the ventilation air rate needed for any of these cases, but how to implement the rates, via system types and equipment choices, is a design decision of yours.

EXAMPLE 6.4

What is the ventilation air-flow rate ($\dot{V}_{tot}$) needed for a 120 ft^2 smoking-allowed single-person office? One occupant and perfect mixing are assumed.

Solution:

Because the space is assumed to be well mixed, the base ventilation air-flow rate ($\dot{V}_{vrp}$) from Table 6.1 of Standard 62.1-2004 is

$$\dot{V}_{vrp} = 1\ p \cdot 5\frac{CFM}{p} + 120\ ft^2 \cdot 0.06\frac{CFM}{ft^2} = 12.2\ CFM \approx 13\ CFM$$

Next, the flow rate needed for diluting the ETS must be added to this base value. The sole occupant is assumed to be a smoker, so $X_{sm} = 1.0$. Because this is a work area, the occupant will be assumed to be long term and thus adapted. Table 4.1a then provides the volumes of air needed to dilute the smoke as $V_{cig,sm} \approx 1,100$ ft^3/cig. Using equation 4.19 with no nonsmokers present, the volume of dilution air per cigarette is then

$$V_{cig} = 1.0 \cdot 1,100 = 1,100\frac{ft^3}{cig}$$

Table 4.2 provides guidance on the smoking rate, and a value of $\dot{R}_{sm} = 0.6$ cig/p$_{sm}$·h is suggested for this occupancy. As the total occupancy (P_{tot}) is just 1 p, the design smoking density is, therefore,

$$\dot{D}_{cig} = \frac{1.0 \frac{p_{sm}}{p} \cdot 1p \cdot 0.6 \frac{cig}{p_{sm} \bullet h}}{120 \, ft^2} = 0.005 \frac{cig}{h \bullet ft^2}$$

As the space is well mixed, the contaminant removal effectiveness, E_{cr}, is assumed to be 1.0. From equation 4.17a, the extra ventilation air needed to dilute the ETS to acceptable odor and irritant levels is then

$$\dot{V}_{ets} = \frac{0.005 \frac{cig}{h \bullet ft^2} \cdot 1,100 \frac{ft^3}{cig} \cdot 120 \, ft^2}{60 \frac{min}{h} \cdot 1.0} \approx 11 \, CFM$$

The total ventilation air-flow rate required for this space, from equation 4.16, is then

$$\dot{V}_{tot} = 13 \, CFM + 11 \, CFM = 24 \, CFM$$

or about 185% of the base requirement for a similar nonsmoking space.

When this calculation is done for an unadapted occupant, a total ventilation air-flow rate of 27 CFM/p (13 LPS/p) is found. This represents 208% of the base, non-ETS ventilation air requirement from Standard 62's Ventilation Rate Procedure.

This example was for a single, adapted smoker in the office. Many non-smoking visitors to this office would likely find the air to be unacceptable. If this were a salesperson's office in an automotive dealership, for example, designing for having an unadapted visitor present as well may be more appropriate.

EXAMPLE 6.5

The location of the smoker can be predicted to be seated at the office's desk, and local exhaust is specified by the designer. If all else from Example 6.4 is the same, what happens if the contaminant removal effectiveness (E_{cr}) is now greater than 1.0 due to better than perfect mixing performance of the exhaust system?

From Example 6.4, $\dot{V}_{vrp}$ is 13 CFM and is unchanged because local exhaust is being used, rather than displacement air flow, to increase E_{cr}. $\dot{V}_{ets}$, however, varies as E_{cr} increases from 1.0. For values of E_{cr} from 1.0, for well-mixed air conditions, to 4.0, for excellent ETS removal before mixing, the new total ventilation air-flow rates, $\dot{V}_{tot} = \dot{V}_{vrp} + \dot{V}_{ets}$, are:

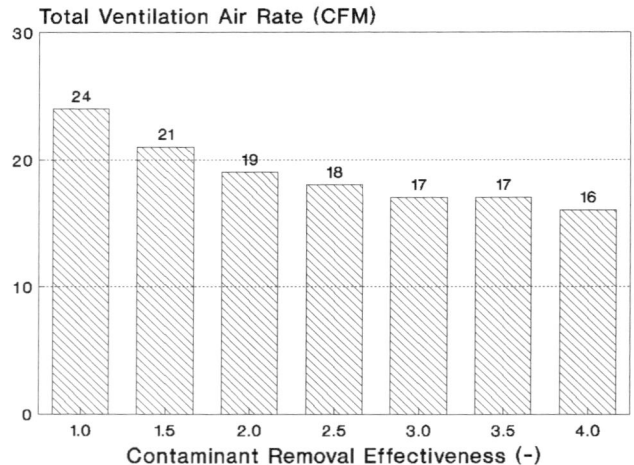

The results show diminishing returns for high E_{cr} values—above ~2.5 yields little reduction in ventilation air-flow rate and thus low potential for more energy savings. As construction and possibly operating costs are likely to increase with higher E_{cr} values, the incremental economics need to be evaluated and may show that very high E_{cr} values are not justified. However, the same analysis may show that increasing the ventilation system performance from $E_{cr} = 1.0$ to just 1.5 or 2.0, for example, can yield significant savings and may be economically feasible.

In this example local exhaust was assumed, so $\dot{V}_{vrp}$ was unchanged from the previous example. ANSI/ASHRAE Standard 62.1-2004 provides an adjustment factor if displacement flow is used instead, and in that case the previous $\dot{V}_{tot}$ would be slightly lower.

6.1.3. Conference Rooms

Designing HVAC systems for conference and similar rooms, as shown in Figure 6.2, is particularly challenging. Their occupancy is often zero for

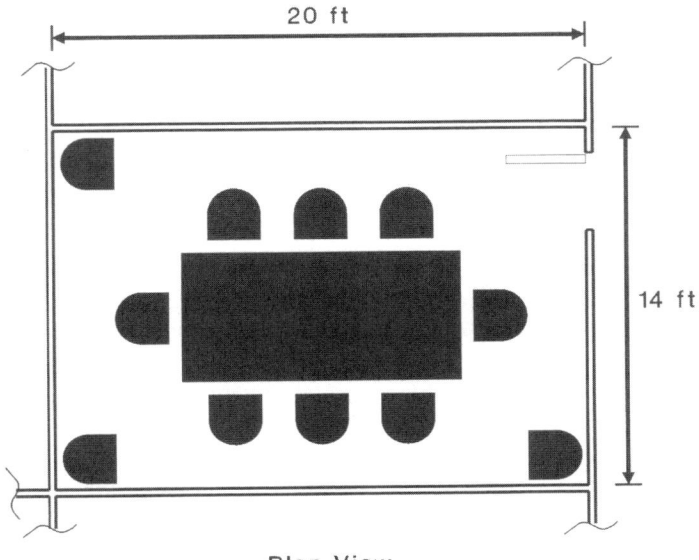

Figure 6.2. A conference room often has a central, large table, around which smokers and others will likely be seated. When possible, local, overhead exhaust should be used to remove the ETS quickly, but this may or may not be possible over this room's table.

most of a typical business day, but then full occupancy occurs suddenly, and later, just as quickly, the occupancy typically drops back to zero. Both thermal and IAQ complaints are frequently generated because either their HVAC systems don't have the capacity needed for the peak loads and/or the control systems can't respond quickly enough. As the occupancy is highly variable in a conference room, a good, adjustable-capacity HVAC system seems appropriate so that the space is not overcooled or overventilated at most hours, but then can supply large quantities of conditioning and ventilating air when needed. Fast-response occupancy sensors are likely needed as part of the zones' controls systems.

Addendum *o* for Standard 62-2001, adopted in 2003, changed the base VRP rates for "Offices: Conference Rooms, Lounges" from 20 to 15 CFM/p (10 to 8 LPS/p). This change was made because the standard's former Table 2 no longer includes an allowance for ETS. In 2004, Addendum *n* changed the rates yet again. However, if a conference room is designated as an ETS area, then more ventilation air will be needed thar these base VRP rates if additivity is assumed. If the room is surrounded by other ETS spaces, then dilution ventilation may be appropriate; if not, then local exhaust, blocking off any returns, and operating the room at a negative air pressure is likely best. Most of the smokers will probably be seated around a large, centrally located table; it may be possible to design the exhaust to increase somewhat the ETS contaminant removal effectiveness.

EXAMPLE 6.6

What is the ventilation air-flow rate ($\dot{V}_{tot}$) needed for a 200 ft^2 smoking-allowed conference room? Perfect mixing of the room's air via ceiling-based air terminals is assumed.

Solution:

The first step is to find the base ventilation air-flow rate from Table 6.1 of Standard 62.1-2004. For this open-plan office space, a value of $V_{vrp}/p = 6$ CFM/p is found from Table 6.1's combined "conference/meeting room" entry. The table's estimate for the occupancy in the space is 50 people per 1000 ft^2, so the Ventilation Rate Procedure's base flow rate ($\dot{V}_{vrp}$) is

$$\dot{V}_{vrp} = 200\ ft^2 \cdot \frac{50\ p}{1,000\ ft^2} \cdot 6\frac{CFM}{p} = 60\ CFM$$

This 60 CFM is the base, non-ETS ventilation air-flow rate required to ventilate the space for all other contaminants. The flow rate needed for diluting the ETS must be found and added to this base value.

The next step is to estimate the percent of occupants who are smokers. In this case a value of 20% smokers is selected from Table 4.2 of this book, which gives a range of 0.2 to 0.25 for X_{sm} for "all other occupancies." The occupants in this space will be assumed to be long term and thus adapted. Table 4.1a then provides the volumes of air needed to dilute the smoke as $V_{cig,ns} \approx 3,900$ ft^3/cig and $V_{cig,sm} \approx 1,100$ ft^3/cig. Using equation 4.19, the adjusted volume of dilution air per cigarette is then

$$V_{cig} = 0.2 \cdot 1,100 + (1 - 0.2) \cdot 3,900 = 3,340 \frac{ft^3}{cig}$$

Table 4.2 provides guidance on the smoking rate, and a value of $\dot{R}_{sm} = 0.6$ cig/p_{sm}·h is suggested for this occupancy. As the total occupancy (P_{tot}) is 10 p, the design smoking density is, therefore,

$$\dot{D}_{cig} = \frac{0.2 \dfrac{p_{sm}}{p} \cdot 10p \cdot 0.6 \dfrac{cig}{p_{sm} \cdot h}}{200 \; ft^2} = 0.006 \frac{cig}{h \cdot ft^2}$$

As the space is well mixed, the contaminant removal effectiveness, E_{cr}, is assumed to be 1.0. From equation 4.17a, the extra ventilation air needed to dilute the ETS to acceptable odor and irritant levels is then

$$\dot{V}_{ets} = \frac{0.006 \dfrac{cig}{h \cdot ft^2} \cdot 3,340 \dfrac{ft^3}{cig} \cdot 200 \; ft^2}{60 \dfrac{min}{h} \cdot 1.0} \approx 67 \; CFM$$

The total ventilation air-flow rate required for this space, from equation 4.16, is then

$$V_{tot} = 60 \; CFM + 67 \; CFM = 127 \; CFM$$

or about 212% of the base requirement for a similar nonsmoking space. For this particular space, the ventilation air-flow rate per person is

$$\frac{\dot{V}_{tot}}{P} = \frac{127 \; CFM}{10 \; p} \approx 13 \frac{CFM}{p}$$

and is thus 7 CFM/p more than the required 6 CFM/p base ventilation rate.

When this calculation is done for unadapted occupants, a total ventilation air-flow rate of 16 CFM/p (8 LPS/p) is found. This represents 267%

of the base, non-ETS ventilation air requirement from Standard 62's Ventilation Rate Procedure as modified by addendum *n* in 2004.

6.2. HOSPITALITY FACILITIES

The hospitality industry is different from many other businesses and is a significant part of the U.S. economy. These restaurants, country clubs, bars. pool halls, cocktail lounges, nightclubs, dance halls, casinos, bowling alleys, health clubs, salons, theaters, and similar facilities cater to the nutritional, social, and entertainment needs of people. Often these occupancies have a higher percentage of smokers than in the general, surrounding populations, but some have lower. Most of these businesses are oper to the public, and encourage frequent visits and associated spending, but some have restricted memberships or allow access to adults only, for example. Local smoking-control ordinances often apply to public buildings. but these smoking regulations may not apply to private clubs. Check the applicable local and state regulations carefully; they may change with time as laws are enacted and rulings made.

Due to some historic catastrophic fires and significant loss of life, smoking in public theaters is likely prohibited. But the lobby area, if allowed by codes or ordinances, may be designated as an ETS area, or, alternatively, a smoking lounge or convenient outdoor smoking area might be provided for employees and customers who smoke.

A valid concern in designing for hospitality facilities and other businesses is the comfort and health of employees. If smoking is allowed, an employee will likely have a long exposure to ETS and any other contaminants that are present. Designing to meet the needs of the adapted employees may be more restrictive than for any unadapted, short-term visitors. If ETS exposure limits and durations are eventually published by the appropriate authorities, similar to that already done for many other contaminants, then the ventilation system's performance will likely need evaluation for both short-term visitors' and long-term employees' exposures; the more restrictive of the two should guide the policy decisions and ventilation systems' designs.

As thermal loads and occupancies vary significantly in the hospitality industry, many different types of HVAC systems are employed. Some are VAV, but often CAV systems are used where ventilation is already of concern (e.g., in bars and restaurants). Your ever-growing design skills will be tested as you try to create comfortable indoor environments and acceptable IAQ, as well as to optimize energy use for these businesses.

6.2.1. Restaurants

A simple restaurant is composed of a kitchen, a food storage area, restrooms, a waiting area, and the main seating space (ASHRAE 2003b, ch. 3). Only employees are typically in the kitchen and storage areas, but both the employees and many customers are in the more public areas of the restaurant. The grills and ovens in the kitchen normally require vast amounts of exhaust air to control the fumes and airborne grease that they produce. As such, the kitchen is often running at a substantial negative air pressure as compared with the seating area and likely to the outside too. Significant discussion of kitchen ventilation and related issues appears in the *ASHRAE Handbook* (ASHRAE 2003b, ch. 31) and elsewhere. At least for sanitation reasons, smoking will hopefully not be allowed in the kitchen. When a door from the kitchen goes outdoors, employees often use it for smoking breaks, however, and ETS can be drawn back into the kitchen. Where possible, providing an enclosed and ventilated mini-smoking break-room at this exit would be of assistance to the employees during inclement weather and should reduce the backflow of smoke into the kitchen.

All or part of the seating area of a restaurant may be a smoking-allowed area by the owner/operator. But providing a separate "no smoking area" can help keep more customers happy, as long as ETS is not apparent in that space. Merely separating the seating, but still using a common highly recirculating and thus mixing, low ventilation rate, low-effectiveness air-cleaning HVAC system to serve both areas, is inappropriate at best. Using a separate ventilation system for the smoking-allowed area and carefully setting the pressurization so that a degree of horizontal displacement flow is created that reduces ETS migration is advisable. A physical floor-to-ceiling barrier, or other engineered method of providing separation such as a low-mixing and thus high contaminant removal effectiveness horizontal displacement flow, may be required to meet the design objectives or to comply with codes or ordinances.

As the kitchens, restrooms, and ETS areas of restaurants have high exhaust air requirements, a large amount of makeup air will need to be admitted and conditioned. Supplying this makeup air to the no-smoking seat_ng area alone may be possible, but the transfer air velocities may be too high. Using heat recovery to precondition the makeup air should be considered (Jenkins et al. 2001). Exhaust air from kitchen hoods, however, usually contains grease and contaminants that can clog heat transfer media, and grease fires are significant hazards (ASHRAE 2003b, ch. 31).

Ventilation, filtration, and heat-recovery principles for handling ETS should not only be applied to proposed restaurants, but also potentially to exis:ing ones undergoing major renovations or when addressing specific IAQ complaints. But for restaurants that operate on slim profit margins, such as those that are smaller and family-owned, meeting the initial and operating costs of the ventilation improvements may be problematic or ever prohibitive. For these owners, an indoor smoking ban may be the only cost-effective ETS control technique.

EXAMPLE 6.7

What is the ventilation air-flow rate ($\dot{V}_{tot}$) needed for a 3,000 ft^2 smoking-allowed section of a restaurant's seating area? Perfect mixing of the room's air is assumed.

Solution:

The first step is to find the base ventilation air-flow rate from Standard 62.1. For this open-plan seating area a value of $\dot{V}_{vrp}/p = 10$ CFM/p is found from Standard 62.1-2004 Table 6.1's "restaurant dining rooms" entry. The table's estimate for the occupancy in the space is 70 people per 1,000 ft^2, so the Ventilation Rate Procedure's base flow rate ($\dot{V}_{vrp}$), as modified by addendum *n* in 2004, is

$$\dot{V}_{vrp} = 3,000 \ ft^2 \cdot \frac{70 \ p}{1,000 \ ft^2} \cdot 10 \frac{CFM}{p} = 2,100 \ CFM$$

This 2,100 CFM is the base, non-ETS ventilation air-flow rate required to ventilate the space for all other contaminants. Next, the flow rate needed for diluting the ETS must be found and added to this value.

An estimate of the percentage of occupants who are smokers is now needed. A value of 20% smokers is selected from Table 4.2, which gives a range of 0.2 to 0.25 for X_{sm} for "all other occupancies." The occupants in this space will be assumed to be long term and thus adapted. Table 4.1a then provides the volumes of air needed to dilute the smoke as $V_{cig,ns} \approx$ 3,900 ft³/cig and $V_{cig,sm} \approx 1,100$ ft³/cig. Using equation 4.19, the adjusted volume of dilution air per cigarette is then

$$V_{cig} = 0.2 \cdot 1,100 + (1-0.2) \cdot 3,900 = 3,340 \frac{ft^3}{cig}$$

Table 4.2 provides guidance on the smoking rate, and a value of $\dot{R}_{sm} =$ 0.6 cig/p_{sm}·h is suggested for this occupancy. As the total occupancy (P_{tot}) is 210 p, the design smoking density is, therefore,

$$\dot{D}_{cig} = \frac{0.2 \frac{p_{sm}}{p} \cdot 210p \cdot 0.6 \frac{cig}{p_{sm} \cdot h}}{3,000 \ ft^2} = 0.0084 \ \frac{cig}{h \cdot ft^2}$$

As the space is well mixed, the contaminant removal effectiveness, E_{cr}, is assumed to be 1.0. From equation 4.17a, the extra ventilation air needed to dilute the ETS to acceptable odor and irritant levels is then

$$\dot{V}_{ets} = \frac{0.0084 \frac{cig}{h \cdot ft^2} \cdot 3,340 \frac{ft^3}{cig} \cdot 3,000 \ ft^2}{60 \frac{min}{h} \cdot 1.0} \approx 1,403 \ CFM$$

The total ventilation air-flow rate required for this space, from equation 4.16, is then

$$\dot{V}_{tot} = 2,100 \ CFM + 1,403 \ CFM = 3,503 \ CFM$$

or about 167% of the base requirement for a similar nonsmoking space. For this particular space, the ventilation air-flow rate per person is

$$\frac{\dot{V}_{tot}}{P} = \frac{3,503 \ CFM}{210 \ p} \approx 17 \frac{CFM}{p}$$

and is thus 7 CFM/p more than the required 10 CFM/p base ventilation rate.

When this calculation is performed for unadapted occupants, a total ventilation air-flow rate of 20 CFM/p (10 LPS/p) is found. This represents 200% of the base, non-ETS ventilation air requirement from Standard 62's Ventilation Rate Procedure. When a higher percentage of smokers is used instead, 50% rather than 20%, the rates become 23 CFM/p (adapted) and 28 CFM/p (unadapted).

6.2.2. Bars, Clubs, and Cocktail Lounges

For some reason, social or otherwise, a higher percentage of customers in bars and other drinking establishments tend to be smokers than that of the surrounding population, if smoking is allowed in the facility. In an unscientific way, such people have described their desire "to go out to socialize, have a drink or two, and to smoke," and these businesses normally cater to these wishes. Some smoke-free bars, for example, market themselves to those who want to avoid ETS and may do well. Smoking bans in New York City and elsewhere have caused all businesses to go smoke free. Some other municipal ordinances, for example, restrict smoking to the bar portion of a combination bar/restaurant. If a no-smoking ordinance is not in existence for hospitality facilities in a particular location, the owners often want to allow smoking in all or parts of their businesses.

Typical drinking establishments have at least a bar, sometimes with stools at it, tables for nearby seating, and restrooms. There may or may not be a kitchen, but often food service is provided in larger bars and nightclubs. Bartenders and waiters typically occupy the public portions of bars along with the customers. Patrons move about to socialize, so it may not be practical to designate separate smoking areas, especially in small bars. All of the establishments are thus likely to be smoking allowed, if desired by the owner and allowed by ordinance.

Dilution, with or without ETS-filtered recirculated air, is likely the ventilation approach used in these facilities. But floor-to-ceiling displacement flow should give better than 1.0 contaminant removal effectivenesses. As many of bars are smaller and probably have other occupied spaces adjacent to them, it is likely wise to run the ETS areas at a slight negative pressure. If there is fixed seating, such as at built-in booths or at stools around the bar, vent hoods can help remove ETS from some likely generation locations. Providing conditioned outside or ETS-free transfer air from behind the bar will help reduce the bartenders' exposure but will not eliminate it.

Addendum *o* for Standard 62, adopted in 2003, changed the base VRP rates for "bars, cocktail lounges" from 30 to 20 CFM/p (15 to 10 LPS/p). This change was made because the Standard's Table 2 no longer included an allowance for ETS. In 2004, addendum *n* decreased the VRP rate, now part of 62.1-2004, much further, as shown in the following example.

EXAMPLE 6.8

What is the ventilation air-flow rate ($\dot{V}_{tot}$) needed for a 2,000 ft² smoking-allowed bar? Perfect mixing of the room's air is assumed.

Solution:

The first step is to find the base ventilation air-flow rate from Standard 62.1. For this bar a value of $V_{vrp}/p = 9$ CFM/p is found from Standard 62.1-2004 Table 6.1's "bars, cocktail lounges" entry. The table's estimate for the occupancy in the space is 100 people per 1,000 ft², so the Ventilation Rate Procedure's base flow rate (V_{vrp}) is

$$\dot{V}_{vrp} = 2,000 \ ft^2 \cdot \frac{100 \ p}{1000 \ ft^2} \cdot 9 \frac{CFM}{p} = 1,800 \ CFM$$

This 1,800 CFM is the base, non-ETS ventilation air-flow rate required to ventilate the space for all other contaminants. Next, the flow rate needed for diluting the ETS must be found and added to this base value.

The next step toward this goal is to estimate the percent of occupants who are smokers. A value of 25% is selected from Table 4.2, which gives a range of 0.25 to 0.50 for X_{sm} for "bars, cocktail lounges, casinos, lunch rooms." The occupants in this space will be assumed to be long-term and thus adapted. Table 4.1a then provides the volumes of air needed to dilute the smoke as $V_{cig,ns} \approx 3,900$ ft³/cig and $V_{cig,sm} \approx 1,100$ ft³/cig. Using equation 4.19, the adjusted volume of dilution air per cigarette is then

$$V_{cig} = 0.25 \cdot 1,100 + (1 - 0.25) \cdot 3,900 = 3,200 \frac{ft^3}{cig}$$

Table 4.2 provides guidance on the smoking rate, and values of $\dot{R}_{sm} =$ 1.0 to 2.0 cig/$p_{sm}{\cdot}$h are suggested for this occupancy. The bar in question is expected to have a low rate, so 1.0 is used. As the total occupancy (P_{tot}) is 200 p, the design smoking density is, therefore,

$$\dot{D}_{cig} = \frac{0.25\dfrac{p_{sm}}{p} \cdot 200p \cdot 1.0\dfrac{cig}{p_{sm}\cdot h}}{2,000\ ft^2} = 0.025\frac{cig}{h\cdot ft^2}$$

As the space is well mixed, the contaminant removal effectiveness, E_{cr}, is assumed to be 1.0. From equation 4.17a, the extra ventilation air needed to dilute the ETS to acceptable odor and irritant levels is then

$$\dot{V}_{ets} = \frac{0.025\dfrac{cig}{h\cdot ft^2} \cdot 3,200\dfrac{ft^3}{cig} \cdot 2,000\ ft^2}{60\dfrac{min}{h} \cdot 1.0} \approx 2,667\ CFM$$

The total ventilation air-flow rate required for this space, from equation 4.16, is

$$\dot{V}_{tot} = 1,800\ CFM + 2,667\ CFM = 4,467\ CFM$$

or about 248% of the base requirement for a similar nonsmoking space. For this particular space, the ventilation air-flow rate per person is

$$\frac{\dot{V}_{tot}}{P} = \frac{4,467\ CFM}{200\ p} \approx 23\frac{CFM}{p}$$

and is thus 14 CFM/p more than the required 9 CFM/p base ventilation rate. Note that this 28 CFM/p is just under the 30 CFM/p listed in Table 2 of superseded Standard 62-1989, although the assumptions made for each likely differ somewhat.

When this calculation is redone for unadapted occupants, a total ventilation air-flow rate of 28 CFM/p (14 LPS/p) is found. This represents 311% of the base, non-ETS ventilation air requirement from Standard 62.1's Ventilation Rate Procedure. When a higher percent of smokers is used instead, 50% rather than 25%, and a higher frequency of smoking is expected, 2.0 rather than 1.0 cigarettes per smoker per hour, the rates become 51 CFM/p (adapted) and 68 CFM/p (unadapted).

EXAMPLE 6.9

What is the ventilation air-flow rate ($\dot{V}_{tot}$) needed for a 2,000 ft² smoking-allowed bar? Instead of designing the ventilation system for perfect mixing, a degree of displacement flow is created via a system with a demonstrated contaminant removal effectiveness of 2.0.

Solution:

The process is the same as for example 6.8, except the contaminant removal effectiveness is increased from 1.0 to 2.0. So, the Ventilation Rate Procedure's base flow rate ($\dot{V}_{vrp}$), without adjusting for displacement flow to be conservative, is

$$\dot{V}_{vrp} = 2,000 \ ft^2 \cdot \frac{100 \ p}{1,000 \ ft^2} \cdot 9 \frac{CFM}{p} = 1,800 \ CFM$$

The occupants in this space will be assumed to be long term and thus adapted. Using equation 4.19, the adjusted volume of dilution air per cigarette is then

$$V_{cig} = 0.25 \cdot 1,100 + (1 - 0.25) \cdot 3,900 = 3,200 \frac{ft^3}{cig}$$

Using a smoking rate of $\dot{R}_{sm} = 1.0$ cig/p$_{sm}$·h, since the bar in question is expected to have a low rate, and the total design occupancy (P_{tot}) is 200 p, the design smoking density is

$$\dot{D}_{cig} = \frac{0.25 \frac{p_{sm}}{p} \cdot 200 \, p \cdot 1.0 \frac{cig}{p_{sm} \cdot h}}{2,000 \ ft^2} = 0.025 \frac{cig}{h \cdot ft^2}$$

As the ventilation and exhaust systems are designed to provide a significant degree of displacement flow, the known contaminant removal effectiveness, E_{cr}, is 2.0. From equation 4.17a, the extra ventilation air needed to dilute the ETS to acceptable odor and irritant levels is then

$$\dot{V}_{ets} = \frac{0.025 \frac{cig}{h \bullet ft^2} \bullet 3200 \frac{ft^3}{cig} \bullet 2,000 \ ft^2}{60 \frac{min}{h} \bullet 2.0} \approx 1,334 \ CFM$$

The total ventilation air-flow rate required, from equation 4.16, is

$$\dot{V}_{tot} = 1,800 \ CFM + 1,334 \ CFM = 3,134 \ CFM$$

or about 174% of the base requirement for a similar perfectly mixed non-smoking space. For this particular space, the ventilation air-flow rate per adapted person is

$$\frac{\dot{V}_{tot}}{P} = \frac{3,134 \ CFM}{200 \ p} \approx 16 \frac{CFM}{p}$$

and is thus 7 CFM/p more than the required 9 CFM/p base ventilation rate.

6.2.3. Casinos

Large, modern casinos are multipurpose buildings; they often include not only gaming areas, but also typically restaurants, theaters, retail shops, arcades, and frequently other entertainment attractions such as pools or amusement rides. These casinos typically operate at all hours and on every day of the year, and peak occupancy can occur on holidays. Because in most casinos the occupancy is highly variable, and often the casinos are well funded, cutting-edge demand, controlled ventilation systems can normally be employed. The owners are frequently willing to install extra capacity to help ensure occupant comfort and, to varying degrees, are willing to experiment with new approaches.

For the purposes of this section, only the gaming portions of a casino are considered. These gambling areas often have noticeably different loads than other portions of the buildings. For example, a large portion of the gaming floor area will likely be dedicated to slot machines, but the classic mechanical machines have now almost completely given way to electro-mechanical slots, video poker, and other electronic game machines that generate significant sensible heat loads. These machines, and their users, are grouped closely together in rows. An occupant will spend most of his or her time sitting on a stool or standing at a machine, but also will spend some time roaming the aisles looking for the next "lucky" machine.

Another significant portion of the gaming area is devoted to the traditional table games, such as blackjack, poker, craps, and roulette. Occupants will mostly be gathered around the tables because they typically already have their favorite games. Dealers will be stationed at some or all of the tables, whether or not customers are currently there.

Other gambling areas often have seating areas facing large displays, such as for sports betting and random-drawing games. Unobstructed viewing of the screens is essential, so having low-hanging vent hoods over the seating areas is not likely feasible.

All of the large gaming areas typically have very high ceilings, so the room air volume is significant and can dilute brief surges in smoke production. ETS should be allowed to rise to this considerable headroom above the occupied zone and be exhausted from there. For security, many monitoring cameras are normally mounted in these high ceilings, and their views should not be restricted by chases, ductwork, vent hoods, or other equipment. Because the large air volumes present in many casinos and atria can store surges, designers may choose to have the ventilation systems' responses "lag." More information on leading and lagging times can be found in Standard 62.1-2004 (ANSI/ASHRAE 2004a) and its new user's guide.

Due to the free movement of occupants, dilution ventilation, with or without ETS-cleaned recirculated air, seems most appropriate for typical large gaming spaces unless floor-to-ceiling displacement flow is possible. Addendum *o* for Standard 62, adopted in 2003, changed the base ventilation rate procedure (VRP) values for well-mixed "gambling casinos" from 30 to 20 CFM/p (15 to 10 LPS/p). This change was made because Standard 62-2001's Table 2 no longer included an allowance for ETS. Addendum *n*, published in 2004, then reduced the non-ETS base rate much further, as shown in the following example.

EXAMPLE 6.10

What is the ventilation air-flow rate ($\dot{V}_{tot}$) needed for a casino's 20,000 ft^2 smoking-allowed gaming floor? Perfect mixing of the room's air via ceiling-based air terminals is assumed.

Solution:

The first step is to find the base ventilation air-flow rate from Standard 62.1. For this open space a value of $\dot{V}_{vrp}/p = 9$ CFM/p is found from Standard 62.1-2004 Table 6.1's "gambling casinos" entry, which was modified by 62's addendum o in 2003 and then again by addendum n in 2004. The table's estimate for the peak normal occupancy in the space is 120 people per 1,000 ft^2, so the Ventilation Rate Procedure's base flow rate ($\dot{V}_{vrp}$) is

$$\dot{V}vrp = 20,000 \ ft^2 \cdot \frac{120p}{1,000 \ ft^2} \cdot 9\frac{CFM}{p} = 21,600 \ CFM$$

This 21,600 CFM is the base, non-ETS ventilation air-flow rate required to ventilate the space for all other contaminants. It is very high, so demand-controlled ventilation and heat recovery should be investigated, as should evaporative cooling if in a dry location. The flow rate needed for diluting the ETS must be found and added to this base value if additivity is assumed.

The next step is to estimate the percent of occupants who are smokers. A value of 25% smokers is selected from Table 4.2, which gives a range of 0.25 to 0.50 for X_{sm} for "bars, cocktail lounges, casinos, lunch rooms." The occupants in this space will be assumed to be long-term and thus adapted. Table 4.1a then provides the volumes of air needed to dilute the smoke as $V_{cig,ns} \approx 3,900$ ft^3/cig and $V_{cig,sm} \approx 1,100$ ft^3/cig. Using equation 4.19, the adjusted volume of dilution air per cigarette is then

$$V_{cig} = 0.25 \cdot 1,100 + (1 - 0.25) \cdot 3,900 = 3,200 \ \frac{ft^3}{cig}$$

Table 4.2 provides guidance on the smoking rate, and values of $\dot{R}_{sm} = 1.0$ to 2.0 cig/p_{sm}·h are suggested for this occupancy. The casino floor in question is expected to have a low rate, so 1.0 is used. As the total occupancy (P_{tot}) is 2,400 p, the design smoking density is, therefore,

$$\dot{D}_{cig} = \frac{0.25\dfrac{p_{sm}}{p} \bullet 2,400p \bullet 1.0\dfrac{cig}{p_{sm} \bullet h}}{20,000 \ ft^2} = 0.03\frac{cig}{h \bullet ft^2}$$

As the space is well mixed, the contaminant removal effectiveness, E_{cr}, is assumed to be 1.0. From equation 4.17a, the extra ventilation air needed to dilute the ETS to acceptable odor and irritant levels is then

$$\dot{V}_{ets} = \frac{0.03\dfrac{cig}{h \bullet ft^2} \bullet 3,200\dfrac{ft^3}{cig} \bullet 20,000 \ ft^2}{60\dfrac{min}{h} \bullet 1.0} \approx 32,000 \ CFM$$

Note that if a displacement system with a contaminant removal effectiveness of 2.0 were used instead, this portion would be halved. The total ventilation air-flow rate required for this space, assuming dilution ventilation with $E_{cr} = 1.0$, from equation 4.16, is

$$\dot{V}_{tot} = 21,600 \ CFM + 32,000 \ CFM = 53,600 \ CFM$$

or about 248% of the base requirement for a similar nonsmoking space. For this particular space, the ventilation air-flow rate per person is

$$\frac{\dot{V}_{tot}}{P} = \frac{53,600 \ CFM}{2400 \ p} \approx 23\frac{CFM}{p}$$

and is thus 14 CFM/p more than the required 9 CFM/p base ventilation rate.

When this calculation is made for unadapted occupants instead, a total ventilation air-flow rate of 28 CFM/p (14 LPS/p) is found. This represents 311% of the base, non-ETS ventilation air requirement from Standard 62.1's Ventilation Rate Procedure. When a higher percentage of smokers is used instead, 50% rather than 20%, and a greater frequency of smoking is expected, 2.0 rather than 1.0 cigarettes per smoker per hour, the rates become 51 CFM/p (adapted) and 68 CFM/p (unadapted).

6.3. SMOKING LOUNGES

Smoking break-rooms or lounges received some public interest and research attention in recent years (e.g., Straub et al. 1993; Alevantis et al. 2003; Jadud and Rock 1993). They have been promoted as an alternative to complete indoor smoking bans in public buildings and workplaces. Typically these rooms are small, enclosed spaces that allow smokers to segregate themselves from others when they desire to smoke (RJR 1992) and thus can achieve excellent odor and irritation control in the surrounding spaces if well designed and operated. The ETS is exhausted to the outdoors rather than being allowed to mix with the rest of the building's indoor air. After using a conveniently located break-room, smokers should then be able to return quickly to their work or other activity. As such, smoking lounges should be distributed so that users can move rapidly to and from them as needed.

One common application for smoking lounges is in public transportation facilities, such as airports, ship terminals, and train, subway, and bus stations, where general smoking bans are imposed. As passengers are frequently restricted to the secured areas of these facilities, and thus are not allowed to go outside to smoke, providing smoking rooms help keep these people calm and better suited to potentially long waits; missed boarding times due to "stepping outside" may be reduced too. Smoking is also often banned from the transit vehicles themselves, so allowing that "one last smoke" before boarding can help the smokers cope.

The other most-common application for these smoking spaces is as break-rooms in workplaces. Instead of making workers go outside, spaces with appropriate ventilation considerations are provided to allow smokers to take breaks indoors as needed. A substantial benefit is that by placing rooms closer to the workplaces, transit times each way can be reduced, and thus workers will be back at their workstations sooner. These break-rooms also should reduce the number of smokers being outdoors near building entrances, so complaints from passers-by should be reduced too.

Design advice varies on whether smoking lounges should be optimized so that work can be performed there, such as by installing phones and computers, or whether they should be made as stark and otherwise uncomfortable as possible to encourage employees to return to their workstations quickly. In either case, selecting surfaces and furnishings that are durable, fire-resistant, nonabsorbing, and cleanable is important. Some have suggested that smoking rooms should be located no more than three floors apart and near stairways to minimize elevator use and that heavy foot-traffic near the smoking lounges should be considered so as not to adversely affect other business activities.

Smoking lounges can often be designed using existing rooms. Air supply from already installed systems and transfer air are used to displace or dilute the ETS, returns are blocked off and the rooms are well-sealed, and the exhausts are vented directly outdoors to create a negative air pressure (Straub et al. 1993). But because the density of smokers and usage rates may be high in these rooms, more-enhanced engineering measures may be of great benefit. For example, once-through, floor-to-ceiling displacement flow should be considered whenever possible. This may require that a ramp be installed at a room's entryway so that a floor plenum can be created. As discussed in the previous chapter, using a sliding door and an as long as possible entryway will help reduce the migration of ETS out of the room as users enter and leave; existing rooms likely do not have these features, so they will need to be added.

Figure 6.3 shows a conceptual design for a retrofit smoking breakroom. A candidate room near the perimeter of the building is identified, and needed changes are made to allow the space to be reclassified as an ETS area. A short (e.g., six inches or so), raised floor is employed so that transfer air enters the floor plenum through transfer ducts from the hallway and then goes into the space through perforated floor panels. The existing supply and return air ducts are blocked off, and a new exhaust duct is run to a location on the exterior of the building away from air intakes. An exhaust fan is installed at the end of this well-sealed duct and is sized to overcome the pressure losses in the exhaust duct, for transferring of air to the room, and to provide the desired negative pressurization. A continuous duty-rated, rainproof exhaust fan is likely selected; "upblast" fans are preferred over "mushroom" fans so as to better project the ETS away from the building and air intakes. Exhaust grilles are installed in the ceiling, are hard-ducted to the exterior, and the ducts are well sealed. A sliding door is at the end of a short entryway, with air dams near the lowered ceiling, and the last ashtray is placed a few feet farther inside. An opening is made above the door, and gravity one-way (in) dampers are installed in it to allow some transfer air to pass into the entryway when the door is closed. The needed total ventilation air-flow rate can be found through the EDM or other preferred method, and the

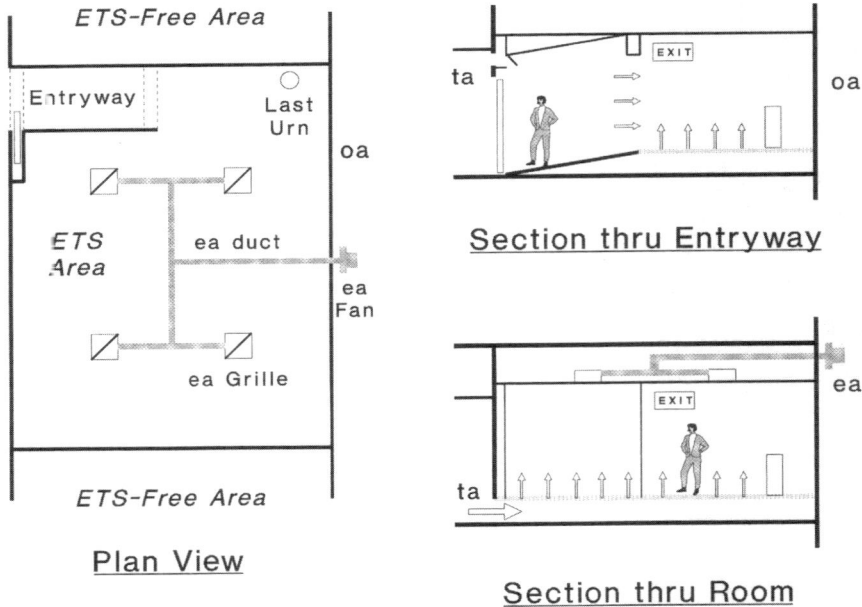

Figure 6.3. A conceptual design to convert an existing small room in a building to a smoking break-room. Transfer air enters under-floor, above the sliding door, and through the doorway when open; thus a degree of displacement flow is created in both the room and the extended entryway. A nearby exhaust fan keeps the space at a negative pressure relative to the neighboring rooms and hall-way. The last ashtray is a heavy or secured urn placed to decrease smoke carry-out during occupants' egress.

designer might choose to use a ventilation effectiveness slightly greater than 1.0 (e.g., 1.2) because this design does provide a degree of floor-to-ceiling displacement flow.

Addendum *o* for Standard 62, adopted in 2003, deleted all entries in the Standard 62-2001 Table 2 for smoking lounges. As the additivity version of the EDM requires a base ventilation rate, a different type of space that models the nonsmoking aspects of a smoking lounge now needs to be identified in what remains of Table 2, now renumbered Table 6.1 via addendum *n* in 2004. Looking through Table 2, "waiting rooms" and "reception areas" seem to be of otherwise similar purposes and have base VRP requirements of 15 CFM/p (7 LPS/p). The previously reported expected occupancy density for smoking lounges was 70 p/1,000 ft^2 (70 p/100 m^2) (ANSI/ASHRAE 1999). Table 6.1 of addendum n and 62.1 has deleted waiting rooms, but does include an entry for "reception areas."

EXAMPLE 6.11

What is the ventilation air-flow rate ($\dot{V}_{tot}$) needed for a 1,000 ft^2 smoking break-room? Perfect mixing of the room's air is assumed.

Solution:

The first step is to find the base ventilation air-flow rate from Standard 62.1-2004. From Standard 62's former Table 2 "waiting room" entry, the estimate for the occupancy in these spaces was 70 people per 1,000 ft^2. From Standard 62.1's Table 6.1 values for reception areas, the Ventilation Rate Procedure's base flow rate ($\dot{V}_{vrp}$) for a nonsmoking break-room is

$$\dot{V}_{vrp} = 1,000 \ ft^2 \cdot \frac{70 \ p}{1000 \ ft^2} \cdot 5\frac{CFM}{p} + 0.06\frac{CFM}{ft^2} \cdot 1,000 \ ft^2 = 410 \ CFM$$

This 410 CFM is the base, non-ETS ventilation air-flow rate required to ventilate the space for all other contaminants and is about 6 CFM/p for this particular space. Additionally, the extra flow rate needed for diluting the ETS must be found.

The next step toward this goal is to estimate the percent of occupants who are smokers. A value of 100% is selected from Table 4.2 via X_{sm} for "smoking lounges"; a built-in assumption is that nonsmokers will avoid this space. The occupants will be assumed to be long term and thus adapted. Table 4.1a then provides the volume of air needed to dilute the smoke as $V_{cig,sm} \approx 1,100$ ft^3/cig. Using equation 4.19 with no nonsmokers present, the volume of dilution air per cigarette is then

$$V_{cig} = 1.0 \cdot 1,100 \cdot 1,100 \frac{ft^3}{cig}$$

Table 4.2 provides guidance on the smoking rate, and values of $\dot{R}_{sm} = 3.0$ to 6.0 cig/p$_{sm}$·h are suggested for this occupancy. The smoking break-room in question is expected to have a low rate, so 3.0 is used. As the total occupancy (P_{tot}) is 70 p, the design smoking density is, therefore,

$$\dot{D}_{cig} = \frac{1.0\dfrac{P_{sm}}{p} \cdot 70p \cdot 3.0\dfrac{cig}{P_{sm} \cdot h}}{1000 \ ft^2} = 0.21\frac{cig}{h \cdot ft^2}$$

As the space is well mixed, the contaminant removal effectiveness, E_{cr}, is assumed to be 1.0. From equation 4.17a, the extra ventilation air needed to dilute the ETS to acceptable odor and irritant levels is

$$\dot{V}_{ets} = \frac{0.21\dfrac{cig}{h \cdot ft^2} \cdot 1{,}100\dfrac{ft^3}{cig} \cdot 1{,}000 \ ft^2}{60\dfrac{min}{h} \cdot 1.0} \approx 3850 \ CFM$$

Note that if a ventilation system were designed to have a contaminant removal effectiveness significantly higher than 1.0, this air-flow rate would be considerably lower. But still assuming perfect mixing, the total ventilation air-flow rate required for this space, from equation 4.16, is

$$\dot{V}_{tot} = 410 \ CFM + 3{,}850 \ CFM = 4{,}260 \ CFM$$

or about 1,039% of the base requirement for a similar nonsmoking space. For this particular space, the ventilation air-flow rate per person is

$$\frac{\dot{V}_{tot}}{P} = \frac{4{,}260 \ CFM}{70 \ p} \approx 61\frac{CFM}{p}$$

and is thus 55 CFM/p more than the required 6 CFM/p base ventilation rate. This 61 CFM/p compares well with the findings of Straub et al. (1993).

When this calculation is done for unadapted occupants, a total ventilation air-flow rate of 76 CFM/p (36 LPS/p) is found. This represents 1,298% of the base, non-ETS ventilation air requirement from Standard 62.1's Ventilation Rate Procedure. When a higher frequency of smoking is expected, 6.0 rather than 3.0 cigarettes per smoker per hour, the rates become 116 CFM/p (adapted) and 146 CFM/p (unadapted). As the occupancy of smoking lounges is typically highly variable, demand-controlled ventilation, with suitable vent-out periods after occupancy, should be investigated to save energy. Displacement ventilation with high contaminant removal effectiveness would greatly reduce energy use even further.

6.4. PRISONS

Prison populations are another demographic that typically have a higher percentage of smokers than that of the general public. Some jails and prisons now ban smoking, but many still allow tobacco products; they may limit their use to individual cells, cellblocks, recreational areas, or other specific locations, however. Often smoking is not allowed in individual cells, and designated smoking areas are defined elsewhere.

Where smoking is allowed in cells, ventilating them is an interesting problem, because nonsmokers may be forced to reside with smokers. In low- to mid-level security facilities, special air terminals may be present within jail cells, but in high-security prisons, their presence is often considered too high a risk. Air is commonly blown into the cells from a duct running just outside and above the cells' bars; air and contaminants are then transferred back into the hallway by momentum for return to the HVAC system or for exhaust. Dilution ventilation is likely the only option in such situations, but, where possible, a ceiling-mounted exhaust grille in the back of a cell might help improve the ETS removal process slightly. The following example calculation assumes two inmates in a small cell, and one is a smoker. As prison overcrowding is currently a significant concern in the United States, the HVAC designer may want to increase the occupancy per cell but also provide adjustability if the prison population decreases in the future.

EXAMPLE 6.12

What is the ventilation air-flow rate ($\dot{V}_{tot}$) needed for a 100-ft^2 jail cell with two prisoners, one of whom is a smoker? Perfect mixing of the cell's air is assumed.

Solution:

The first step is to find the base ventilation air-flow rate from Standard 62.1. For this room Standard 62.1-2004 Table 6.1's "cells" entry estimates the occupancy in these spaces is 30 people per 1,000 ft^2, which is somewhat higher than the known 2 p in this particular 100-ft^2 cell. Using Table 6.1's Ventilation Rate Procedure values the base flow rate (V_{vrp}) for this two-person nonsmoking prison cell is

$$\dot{V}_{vrp} = 2p \cdot 5\frac{CFM}{p} + 100 \ ft^2 \cdot 0.12\frac{CFM}{ft^2} = 22 \ CFM$$

or about 11 CFM/p. This 22 CFM is the base, non-ETS ventilation air-flow rate required to ventilate the space for all other contaminants. Assuming additivity, next the flow rate needed for diluting the ETS must be found and added to this base value.

The percent of occupants who are smokers is now needed. Table 4.2 does not list jail cells, and the "all others" X_{sm} of 0.20 to 0.25 seems low. After talking with the client the designer estimates, on average, 50% of the occupants will be smokers. The occupants will be assumed to be long-term inmates and thus adapted. Table 4.1a then provides the volumes of air needed to dilute the smoke as $V_{cig,ns} \approx 3,900$ ft^3/cig and $V_{cig,sm} \approx 1,100$ ft^3/cig. Using equation 4.19, the adjusted volume of dilution air per cigarette is then

$$V_{cig} = 0.5 \cdot 1,100 + 0.5 \cdot 3,900 = 2,500\frac{ft^3}{cig}$$

Table 4.2 provides guidance on the smoking rate, and a value of $\dot{R}_{sm} = 0.6$ cig/p$_{sm}$·h is suggested for "all other" occupancies; again, the designer thinks this rate is low for prisoners and decides to use 1.0. As the total occupancy (P_{tot}) of the cell is 2 p, the design smoking density is, therefore,

$$\dot{D}_{cig} = \frac{0.5\frac{p_{sm}}{p} \cdot 2p \cdot 1.0\frac{cig}{p_{sm} \cdot h}}{100 \ ft^2} = 0.01\frac{cig}{h \cdot ft^2}$$

As the space is well mixed, the contaminant removal effectiveness, E_{cr}, is assumed to be 1.0. From equation 4.17a, the extra ventilation air needed to dilute the ETS to acceptable odor and irritant levels is then

$$\dot{V}_{ets} = \frac{0.01 \frac{cig}{h \cdot ft^2} \cdot 2,500 \frac{ft^3}{cig} \cdot 100 \, ft^2}{60 \frac{min}{h} \cdot 1.0} \approx 42 \, CFM$$

The total ventilation air-flow rate required for this space, from equation 4.16, is

$$\dot{V}_{tot} = 22 \, CFM + 42 \, CFM = 64 \, CFM$$

or about 291% of the base requirement for a similar nonsmoking space. For this particular space, the ventilation air-flow rate per person is

$$\frac{\dot{V}_{tot}}{P} = \frac{64 \, CFM}{2p} = 32 \frac{CFM}{p}$$

and is thus 21 CFM/p more than the required 11 CFM/p base ventilation rate.

When this calculation is done for unadapted occupants, a total ventilation air-flow rate of 41 CFM/p (20 LPS/p) is found. This represents 373% of the base, non-ETS ventilation air requirement from Standard 62.1's Ventilation Rate Procedure.

6.5. HOTELS AND MOTELS

Like casinos, large modern hotels have spaces in them for a variety of functions. In addition to the guest rooms, there are often restaurants, bars, clubs, exercise rooms, pools, meeting rooms, ballrooms, lobbies, laundry rooms, parking garages, and other support spaces. Motels, short for "motor hotels," are usually much smaller and simpler but not always. In this section, mainly the guest rooms are considered, but other sections of this chapter and book address the buildings' additional spaces. Hotels and motels can be classified as part of the hospitality industry, but their dominant short-term residential purpose often makes them considered separately.

Most hotels now offer smoking and nonsmoking rooms to accommodate the desires of their guests. But as the percentage of these two groups varies with each night, it is difficult for the managers to always match the room supply to the demand. Some rooms will likely need to swap functions and should be assigned only to occupants who can tolerate the residual ETS. Increased ventilation in these rooms, as much of a flush-out period as possible between occupancies, and extra cleaning may help improve their acceptability. Figure 6.4 shows common room air flows.

Smoking and nonsmoking rooms should not be immediate neighbors, horizontally or vertically, due to the leakage of air between rooms and the ETS that may enter any common hallways. Usually smoking and nonsmoking rooms are arranged by floor, wing, or in large hotel complexes, by building; putting smoking rooms in a separate building achieves the highest degree of separation. Having the smoking rooms in a different wing, with dedicated ventilation systems, is likely the next-best solution, but care is needed to make sure that ETS does not travel through each floor's lobby or the atrium to the nonsmoking wings, for example. Consider any dominant wind directions, since infiltration and transfer air can move ETS horizontally. A well-gasketed door with an automatic closer from the common lobby to a smoking wing's hallway may be necessary, but be sure to consider fire egress requirements. The last but probably most common approach is to designate entire floors of guest rooms as either smoking or nonsmoking; having one or more transitional, reassignable floors in-between is advisable. If smoking and nonsmoking rooms must be immediate neighbors, seal the walls or floors between them well as was described previously and illustrated in Figure 5.3.

The *stack effect* is the movement of air vertically due to buoyancy. When the outdoor air is colder than the indoor air, the relatively warm heated indoor air will rise through a building. In warm weather, the relatively cool air-conditioned indoor air will sink. Figure 6.5 shows that at some point in the building there is a time-varying *neutral pressure level* (*NPL*), and this floor has minimal infiltration or exfiltration due to the stack effect; other floors will have air leakage in or out depending on which way the air is moving vertically.

Considering the stack effect only, in a cold climate it is probably better to place the smoking rooms on the top floors. In a consistently hot

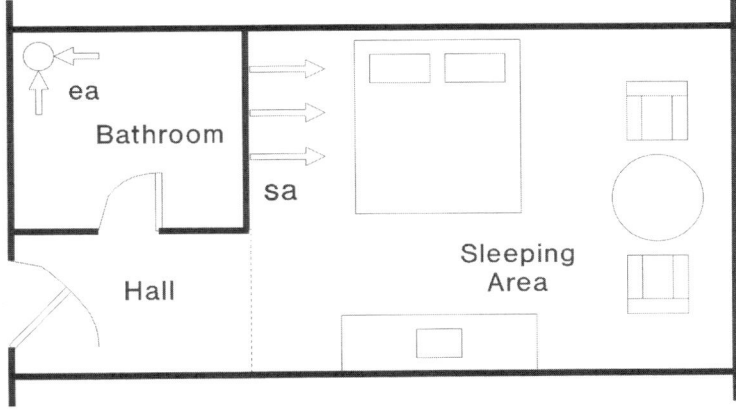

Floor Plan View

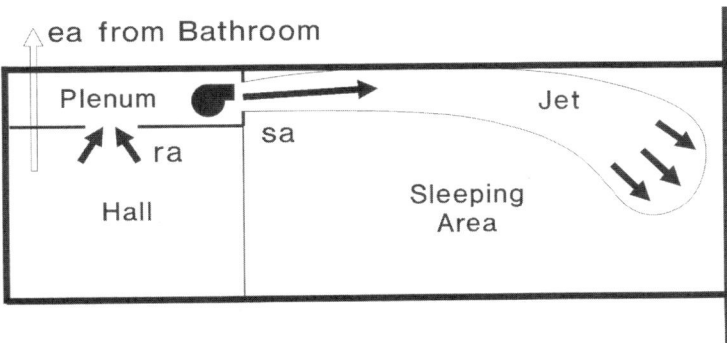

Section View

Figure 6.4. A typical hotel room with a recirculating fan in the plenum over the bathroom and a high-sidewall supply grille to the bedroom. Outside air is often delivered to this mixing plenum, and exhaust air is withdrawn via a bathroom grille normally placed over the shower/bath to help reduce mirror fogging.

climate, the lower floors should be better. Unfortunately, most of the United States has climates that, to varying degrees, are hot in the summer and cold in the winter. One extreme season may be more dominant than the other, such as warm weather in Atlanta or the cold, long winters in Minneapolis. Switching the smoking and nonsmoking floors by season is

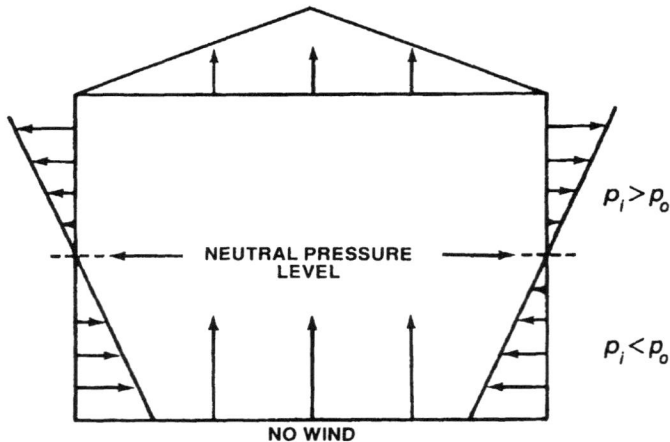

NOTE: Arrows indicate magnitude and direction
of pressure difference

Figure 6.5. Variations in air density cause the stack effect, which is the vertical movement of air through a building. Wind can enhance or reduce this effect, and via infiltration and exfiltration, cause horizontal movement of air through the building as well. With no wind considered, the neutral pressure level can be predicted (ASHRAE 1997, ch. 25.8).

probably not wise due to the absorbed smoke products and the sensitivity of many nonsmokers to lingering odors and irritants.

Sealing the gaps between floors, such as those in pipe, duct, and electrical chases, and installing excellent gaskets on elevator and stair tower doorways, for example, will help minimize the stack effect, as will reducing the paths in the building's skin for infiltration and exfiltration. But floor-by-floor active pressurization and depressurization via advanced HVAC systems may be needed to control the flow of ETS. Or more creative solutions might be employed, such as having an open-air, garden floor between the smoking and nonsmoking levels in a tall building. A pressurized dividing floor of meeting rooms, for example, may also help achieve the needed separation.

The corridors between rows of rooms provide an opportunity to remove some ETS that is moving horizontally out of upwind spaces due to infiltration. However, the thermal loads on hallways tend to be very low, so with conventional design supply air-to-room temperature differences—for example, 20°F (11°C) when in cooling mode—the supply and return air-flow rates for corridors are normally small. Consider using smaller temperature differences, or, if moisture is a significant problem, reheat, so that higher air-flow rates can be moved through halls and thus more ETS removed. For corridors next to known smoking-allowed rooms, also consider using 100% outside air for maximum dilution and exhausting all air removed from these hallways. Active pressurization should also be considered to minimize ETS transfer to halls, as should installing gasketing on all the guest room doors to the corridors.

Computer programs such as NIST's CONTAM model (Dols and Walton 2002) can predict interzonal airflows and pollutant transport, such as the movement of ETS horizontally and/or vertically in a tall building. Both the stack and wind effects can be characterized. But many approximations must be made, such as the selection of the flow coefficients when using the codes. Also, typical weather data are used; uncharacteristically high wind conditions, for example, will yield higher than predicted inter-room ETS transport.

While in the past many hotel and motel rooms had no provisions for forced ventilation air, Standard 62.1 requires a specific rate and thus implies that mechanical ventilation, rather than natural ventilation or infiltration, is needed. Table 6.1 of the Standard 62.1-2004 requires about 11 CFM per person. There is no entry for "baths," but these bathrooms often have exhaust. Upsizing the bathroom exhaust and using it continuously would be a method for removing the ETS from the room. For smoking rooms, the ventilation air-flow rate will need to be increased over the base 11 CFM/p (6 LPS/p) requirement if additivity is assumed. Due to the requirement for negative pressurization in the ETS exhaust duct, however, the fan itself would likely need to be mounted remotely. For tall buildings the exhaust duct would likely serve many rooms and thus require balancing dampers.

It is sometimes unknown which smoking rooms will, at any particular time, be occupied by smokers or nonsmokers. But if a known smoking room is adjacent to a planned non-ETS space, it should be run at a relative negative air pressure. An additional complication is that pressurizations in hotel rooms and across egress doors also need to follow fire code regulations, so consult the project's fire protection engineer on this issue. Sometimes the rooms must be neutral to their hallway to minimize smoke transfer in either direction during a fire event.

EXAMPLE 6.13

What is the general ventilation air-flow rate ($\dot{V}_{tot}$) needed for a 300 ft^2 (not including the restroom) smoking-allowed hotel room with two guests, one of whom is a smoker? Perfect mixing of the room's air is assumed.

Solution:

The first step is to find the base ventilation air-flow rate from Standard 62.1. Only the bedroom area will be considered in this example; the exhaust fan in the restroom may need upsizing too. Table 6.1 of Standard 62.1-2004 gives for the "bedroom/living room" portion of this particular guestroom:

$$\dot{V}_{vrp} = 2p \bullet 5\frac{CFM}{p} + 300 \; ft^2 \bullet 0.06\frac{CFM}{ft^2} = 28 \; CFM$$

or 14 CFM/p. Assuming additivity, the flow rate needed for diluting the ETS must be found and added to this base value.

The next step is to estimate the percent of occupants that are smokers. Table 4.2 does not list hotel rooms, and the "all others" X_{sm} of 0.20 to 0.25 seems low; the designer estimates, on average, that 50% of the occupants of smoking-allowed rooms will be smokers. The occupants will be assumed to be registered guests and thus adapted. Table 4.1a then provides the volumes of air needed to dilute the smoke as $V_{cig,ns} \approx 3,900$ ft^3/cig and $V_{cig,sm} \approx 1,100$ ft^3/cig. Using equation 4.19, the adjusted volume of dilution air per cigarette is then

$$V_{cig} = 0.5 \bullet 1,100 + 0.5 \bullet 3,900 = 2,500\frac{ft^3}{cig}$$

Table 4.2 provides guidance on the smoking rate, and a value of $\dot{R}_{sm} = 0.6$ cig/p$_{sm}$·h is suggested for "all other" occupancies. As the total occupancy (P_{tot}) of the room is 2 p, the design smoking density is, therefore,

$$\dot{D}_{cig} = \frac{0.5\frac{p_{sm}}{p} \bullet 2p \bullet 0.6\frac{cig}{p_{sm} \bullet h}}{300 \; ft^2} = 0.002\frac{cig}{h \bullet ft^2}$$

As the space is well mixed, the contaminant removal effectiveness, E_{cr}, is assumed to be 1.0. From equation 4.17a, the extra ventilation air needed to dilute the ETS to acceptable odor and irritant levels is then

$$\dot{V}_{ets} = \frac{0.002 \frac{cig}{h \cdot ft^2} \cdot 2,500 \frac{ft^3}{cig} \cdot 300 \, ft^2}{60 \frac{min}{h} \cdot 1.0} \approx 25 \, CFM$$

The total ventilation air-flow rate required for this space, from equation 4.16, is

$$\dot{V}_{tot} = 28 \, CFM + 25 \, CFM = 53 \, CFM$$

or about 190% of the base requirement for a similar nonsmoking space. For this particular space, the ventilation air-flow rate per person is

$$\frac{\dot{V}_{tot}}{P} = \frac{53 \, CFM}{2p} \approx 27 \frac{CFM}{p}$$

and is thus 13 CFM/p more than the required 14 CFM/p base ventilation rate when two people occupy the room.

When this calculation is done for unadapted occupants, a total ventilation air-flow rate of 32 CFM/p (15 LPS/p) is found. This represents 229% of the base, non-ETS ventilation air requirement from Standard 62.1's Ventilation Rate Procedure.

6.6. APARTMENTS, CONDOMINIUMS, AND DORMITORIES

Similar to hotels, apartments, condos, and dorms are primarily residential in nature but are typically for longer-term residents. While smaller versions of these residential buildings may have few amenities, larger ones often include various support spaces. High-rise residential buildings normally have lobbies, elevators, central managers' offices, and some include parking garages. If not included in the residential units themselves, there are typically separate and substantial laundry facilities, and people often smoke in them. In some buildings, especially those located in the downtown areas of large cities, restaurants and retail spaces are frequently located in the buildings too, normally at street level. Dormitories may have significant dining, kitchen, study, and entertainment facilities, or these functions may be provided through separate buildings. As these various supporting spaces are covered elsewhere in this book, only the living units are addressed in this section.

Units in dormitories and apartment buildings are rented or leased, so the management typically has some, but often not full-control on where smoking will or will not be allowed. It is wise to designate smoking and nonsmoking units and to place them as was described in the previous section for hotel rooms. Condos, however, are individually owned, so unless strongly described and enforced via the owners' and associations' agreements, or by law, smoking indoors might not be controllable at all. Providing continuous forced ventilation air, running units at neutral pressurization, not recirculating air between units, having excellent air barriers, and sealing penetrations between units should help minimize problems between units' occupants. As with hotels, fire codes may dictate the needed relative pressurizations of rooms, halls, and stair towers, for example. Moisture, via various sources including airborne humidity, needs appropriate consideration too (Harriman et al. 2001)—pressure and temperature differentials across walls can result in condensation.

Table 6.1 of Standard 62.1-2004 includes dormitory occupancies along with the hotels and motels entries, so the example in the previous section is probably most applicable to such buildings' units. Apartments and condos are more likely to fall under Table E-2, formerly 2.3, of the standard that is for "private dwellings, single, multiple." This Table E-2 is significantly different from 6.1 of the standard; it reflects that residential buildings have traditionally relied on infiltration, natural ventilation, and intermittent user-controlled exhaust to meet ventilation needs. A minimum of 15 CFM/p (7.5 LPS/p) is required for the living areas, but in addition a check is needed to be sure that the ventilation air is also at least 0.35 *air changes per hour* (*ACH*). The example that follows demonstrates this needed check. ASHRAE Standard 62.2-2004 has different requirements for low-rise residential buildings and may be applicable to the occupancy of your concern if covered by 62.2's scope.

If the EDM or another per-person method is to be used to determine the ventilation air-flow rate when ETS is present, an estimate of the number of occupants will need to be made. For residential buildings it is often assumed that there are two people for the first bedroom, and then one more person for each additional bedroom. So, for example, a three-bedroom apartment or condo would likely be estimated to have four occupants. Smaller dormitory rooms are typically meant for two people, but larger rooms may be expected to house more.

EXAMPLE 6.14

What is the general ventilation air-flow rate ($\dot{V}_{tot}$) needed for a 500-ft^2 smoking-allowed one-bedroom condominium unit with two residents, one of whom is likely a smoker? Perfect mixing of the condo's air and additivity are assumed.

Solution:

The first step is to find the base ventilation air-flow rate from Standard 62.1. Table E-2 requires that at least 15 CFM/p be provided for "living areas" of residences but also requires at least 0.35 ACH. If this condo unit is assumed to be 8-ft tall, it has a volume of about 4,000 ft^3. The alternative ventilation air requirement is then

$$\dot{V}_{vrp} = 4,000 \ ft^3 \cdot 0.35 \frac{1}{h} = 1,400 \frac{ft^3}{h} \approx 24 \ CFM$$

As this 24 CFM is less than the 30 CFM lower limit, the 15 CFM/p, for the two residents of this particular unit, will be used, since it is the minimum required by Standard 62.1-2004. The flow rate needed for diluting the ETS must be found and added to this base value.

The percent of occupants who are smokers must next be estimated. Table 4.2 does not list condos or apartments, and the "all others" X_{sm} of 0.20 to 0.25 seems low; the designer estimates, on average, 50% of the occupants of smoking-allowed units will be smokers. The occupants will be assumed to be long-term residents and thus adapted. Table 4.1a then provides the volumes of air needed to dilute the smoke as $V_{cig,ns} \approx 3,900$ ft^3/cig and $V_{cig,sm} \approx 1,100$ ft^3/cig. Using equation 4.19, the adjusted volume of dilution air per cigarette is then

$$V_{cig} = 0.5 \cdot 1,100 + 0.5 \cdot 3,900 = 2,500 \frac{ft^3}{cig}$$

Table 4.2 provides guidance on the smoking rate, and a value of $\dot{R}_{sm} = 0.6$ cig/p_{sm}·h is suggested for "all other" occupancies. As the total occupancy (P_{tot}) of the condo is 2 p, the design smoking density is, therefore,

$$\dot{D}_{cig} = \frac{0.5 \dfrac{p_{sm}}{p} \cdot 2p \cdot 0.6 \dfrac{cig}{p_{sm} \cdot h}}{500 \; ft^2} = 0.0012 \frac{cig}{h \cdot ft^2}$$

As the space is well mixed, the contaminant removal effectiveness, E_{cr}, is assumed to be 1.0. From equation 4.17a, the extra ventilation air needed to dilute the ETS to acceptable odor and irritant levels is then

$$\dot{V}_{ets} = \frac{0.0012 \dfrac{cig}{h \cdot ft^2} \cdot 2,500 \dfrac{ft^3}{cig} \cdot 500 \; ft^2}{60 \dfrac{min}{h} \cdot 1.0} = 25 \; CFM$$

The total ventilation air-flow rate required for this space, from equation 4.16, is

$$\dot{V}_{tot} = 30 \; CFM + 25 \; CFM = 55 \; CFM$$

or about 183% of the base requirement for a similar nonsmoking space. For this particular space, the ventilation air-flow rate per person is

$$\frac{\dot{V}_{tot}}{P} = \frac{55 \; CFM}{2p} \approx 28 \frac{CFM}{p}$$

and is thus 13 CFM/p more than the required 15 CFM/p base ventilation rate when two people occupy the unit.

When this calculation is done for unadapted residents, a total ventilation air-flow rate of 33 CFM/p (16 LPS/p) is found. This represents 220% of the base, non-ETS ventilation air requirement from Standard 62.1's Ventilation Rate Procedure.

If a similar calculation is performed for a two-bedroom, 750-ft^2 condominium with three residents, two of whom are smokers, an adapted ventilation air-flow rate of 29 CFM/p (14 LPS/p) is found. For unadapted occupants, the rate is 34 CFM/p (17 LPS/p).

6.7. SUPPORT SPACES

Table 6.1 of ANSI/ASHRAE Standard 62.1-2004 lists Ventilation Rate Procedure values for some other types of occupancies on a per unit floor area basis, i.e., CFM/ft^2 (LPS/m^2), rather than per person. Smoking may be allowed in these hallways, storage rooms, shipping and receiving areas, warehouses, sports arenas, gyms, swimming pools, and similar areas, so additional ventilation air is needed for handling their ETS. If the EDM is used to determine the rates, and since it is a per person method, you will need to estimate the occupancy for each space before calculating the ventilation air requirements.

6.8. USE OF HEAT RECOVERY

Section 5.9.3 described the use of heat recovery for reducing energy consumption and/or increasing the ventilation air-flow rate above a minimum. As recent readings of Standard 62.1-2004 yield, for ETS applications, that something more than the VRP-determined rates must be provided, using heat recovery to increase the ventilation rate above the non-ETS VRP value while not incurring an energy penalty is attractive. In some cases, when the ETS outside air-flow rate is not too far above the base VRP flow rate, adding well-designed, installed, and operated heat recovery can even *reduce* overall energy consumption. The following is a simple steady-state example of the load-savings calculation, but unfortunately much more detailed energy calculations are needed for a real building. Due to the transient thermal mass and weather effects, a full hour-by-hour energy analysis program—for example, TRNSYS, HVACSim+, DOE-2, BLAST, or EnergyPlus—and local weather data must be used to fully evaluate air-side heat recovery.

EXAMPLE 6.15

An existing thermal zone, served by a single-zone CAV air handler, is being converted to a smoking-allowed area. The outside air-flow rate needs to be increased from 200 CFM to 1,000 CFM. The exhaust air-flow rate is the same as the outside air-flow rate both before and after modification The nearby city's winter design condition is 10°F, and in the summer is 96°F DB and 78°F WB. The design supply air temperatures are 100°F and 55°F in the winter and summer, respectively. The indoor air temperature during occupied hours is 72°F year 'round, and, due to near perfect air mixing indoors, the return air temperature is essentially the same. If 50% effective sensible heat-only heat recovery is added as the outside air-flow rate is increased, what is the impact on the peak sensible ventilation loads?

First, the existing peak ventilation loads are found using the sensible heat equation, the outside air-flow rate, and the temperature difference between the outside air and the supply air:

$$\dot{Q}_{s,vent} = 1.1 \cdot CFM \cdot \Delta T = 1.1 \cdot 200 \cdot (100 - 10) = 19,800 \ Btu/h \ \text{(Winter)}$$

$$\dot{Q}_{s,vent} = 1.1 \cdot 200 \cdot (96 - 55) = 9,020 \ Btu/h \cong 0.75 \ tons_R \ \text{(Summer)}$$

Note that these are not the loads imposed by the building or its internal heat gains; they are only the sensible heat rate needed to bring the outside air flow to the supply air conditions.

Next, the new peak ventilation loads, with the higher flow rate and without heat recovery, are found:

$$\dot{Q}_{s,vent} = 1.1 \cdot 1,000 \cdot (100 - 10) = 99,000 \ Btu/h \ \text{(Winter)}$$

$$\dot{Q}_{s,vent} = 1.1 \cdot 1000 \cdot (96 - 55) = 45,100 \ Btu/h \cong 3.8 \ tons_R \ \text{(Summer)}$$

So when the flow rate is increased, the sensible heating capacity of the AHU will need to be increased by about 90,000 Btu/h if no heat recovery is used. As the winter outdoor air is likely dry, humidification will need to be increased, too, if provided. Under summer design conditions, the sensible cooling capacity would need to be increased by about 3 tons$_R$; the latent load, for dehumidification of the extra ventilation air, would increase, too, because the summer design condition for this location is humid, as indicated by the high mean coincident wetbulb temperature.

With the 50% effective heat recovery now added, energy is transferred between the exhaust air being expelled from and the outside air being admitted to the AHU. At winter design conditions, and because the *OA* and *EA* flow rates are the same, the outside air temperature is increased from 10°F to

$$T_{oa,HX} = T_{oa} + \varepsilon_{HX} \bullet (T_{ea} - T_{oa}) = 10 + 0.5 \bullet (72 - 10) = 41°F$$

and in the summer, the 96°F outside air is reduced to

$$T_{oa,HX} = 96 + 0.5 \bullet (72 - 96) = 84°F$$

The new, sensible heat capacity required, when heat recovery is used, is instead

$$\dot{Q}_{s,vent} = 1.1 \bullet 1,000 \bullet (100 - 41) = 64,900 \ Btu/h \ \text{(Winter)}$$

$$\dot{Q}_{s,vent} = 1.1 \bullet 1,000 \bullet (84 - 55) = 31,900 \ Btu/h \cong 2.7 \ tons_R \ \text{(Summer)}$$

So, the needed new, peak sensible heat capacity is 99,000 – 64,900 = 34,100 Btu/h (10 kW) less in the winter and 3.8 – 2.7 = 1.1 tons$_R$ (3.9 kW) less in summer, when 50% effective heat recovery is employed.

This steady-state analysis of the peak sensible loads indicates that significant savings are possible in both winter and summer. However, a full, transient energy analysis is required to predict a typical year's savings. Then an economic analysis would show if the extra cost of the heat-recovery equipment is justified.

6.9. SUMMARY OF EXAMPLES PER PERSON RATES

Table 6.1, in both I-P (6.1a) and SI (6.1b) versions, summarizes many sample calculations made in this chapter with the EDM for various occupancies. The two right-most columns in the table are the total ventilation air-flow rates, including that needed for diluting the ETS. When referring to this table, be aware that each value has a variety of built-in assumptions; users should use the table as a starting point only. The text and examples in this and the preceding chapters describe the occupancies, physics, and other characteristics, and the assumptions used in preparing the table should be compared with that for your current design project. You may, for example, find that an expected smoking rate is much higher than that incorporated in a particular result in Table 6.1, or maybe your occupancy estimate is much lower. You might also decide that the EDM is not the method with which you wish to estimate the extra ventilation air needed to address ETS's odors and irritants; you might, for example, use the method and examples described in Section 4.1 that yield much higher dilutions. As this book was produced at one particular point in time, and the underlying knowledge base will increase, you should also stay current on ETS developments and changes to Standard 62.1, and, of course, look for future, updated editions of this book.

Table 6.1a. Summary of Some Examples and Additional Calculations on a Per Person Basis using the ETS Dilution Method with Specific Assumptions (I-P Units)

Application	Occupancy	X_{sm} (–)	$\dot{R}_{sm}$ (cig/p$_{sm}$·h)	$\dot{V}_{vrp}/p$ (CFM/p)	$\dot{V}_{ets}/p$		$\dot{V}_{tot}/p$	
					Unadapted (CFM/p)	Adapted (CFM/p)	Unadapted (CFM/p)	Adapted (CFM/p)
Office								
Open Plan	5p/1,000 ft²	0.2	0.6	17	10	7	27	24
Single Office	1p	1.0	0.6	13	14	11	27	24
Conference Rm.	50p/1,000 ft²	0.2	0.6	6	10	7	16	13
Hospitality								
Dining Area, Cafeteria (low sm.)	70p/1,000 ft²	0.2	0.6	10	10	7	20	17
Dining Area Cafeteria (high sm.)	70p/1,000 ft²	0.5	0.6	10	18	13	28	23
Bar, Cocktail Lounge (low)	100p/1,000 ft²	0.25	1.0	9	19	14	28	23

Table 6.1a. Summary of Some Examples and Additional Calculations on a Per Person Basis using the ETS Dilution Method with Specific Assumptions (I-P Units) *(continued)*

Application	Occupancy	X_{sm} (–)	$\dot{R}_{sm}$ (cig/p_{sm}·h)	$\dot{V}_{vrp}/p$ (CFM/p)	V_{ets}/p Unadapted (CFM/p)	Adapted (CFM/p)	V_{tot}/p Unadapted (CFM/p)	Adapted (CFM/p)
Hospitality *(continued)*								
Bar, Cocktail Lounge (high)	100p/1,000 ft²	0.5	2.0	9	59	42	68	51
Casino (low)	120p/1,000 ft²	0.25	1.0	9	19	14	28	23
Casino (high)	120p/1,000 ft²	0.5	2.0	9	59	42	68	51
Smoking Lounge								
(low)	70p/1,000 ft²	1.0	3.0	6	70	55	76	61
(high)	70p/1,000 ft²	1.0	6.0	6	140	110	146	116
Prison Cell	2p	0.5	1.0	11	30	21	41	32

Table 6.1a. Summary of Some Examples and Additional Calculations on a Per Person Basis using the ETS Dilution Method with Specific Assumptions (I-P Units) *(continued)*

Application	Occupancy	X_{sm} (–)	$\dot{R}_{sm}$ (cig/p$_{sm}$·h)	$\dot{V}_{vrp}/p$ (CFM/p)	$\dot{V}_{ets}/p$ Unadapted (CFM/p)	Adapted (CFM/p)	$\dot{V}_{tot}/p$ Unadapted (CFM/p)	Adapted (CFM/p)
Hotel or Dorm								
Bedroom	2p	0.5	0.6	14	18	13	32	27
Apartment or Condo Unit								
1 Bedroom	2 p	0.5	0.6	15	18	13	33	28
2 Bedrooms	3 p	0.67	0.6	15	19	14	34	29

Table 6.1b. Summary of Some Examples and Additional Calculations on a Per Person Basis using the ETS Dilution Method with Specific Assumptions (SI Units for flow rates; rounded from I-P equivalents)

Application	Occupancy	X_{sm} (-)	$\dot{R}_{sm}$ (cig/p$_{sm}$·h)	$\dot{V}_{vrp}/p$ (LPS/p)	$\dot{V}_{ets}/p$ Unadapted (LPS/p)	$\dot{V}_{ets}/p$ Adapted (LPS/p)	$\dot{V}_{tot}/p$ Unadapted (LPS/p)	$\dot{V}_{tot}/p$ Adapted (LPS/p)
Office								
Open Plan	5p/100 m^2	0.2	0.6	8	5	3	13	12
Single Office	1p	1.0	0.6	7	7	5	13	12
Conference Rm.	50p/100 m^2	0.2	0.6	3	4	3	8	7
Hospitality								
Dining Area, Cafeteria (low sm.)	70p/100 m^2	0.2	0.6	5	5	3	10	8
Dining Area, Cafeteria (high sm.)	70p/100 m^2	0.5	0.6	5	8	6	14	11
Bar, Cocktail Lounge (low)	100p/100 m^2	0.25	1.0	5	9	7	14	11

Table 6.1b. Summary of Some Examples and Additional Calculations on a Per Person Basis using the ETS Dilution Method with Specific Assumptions (SI Units for flow rates; rounded from I-P equivalents) *(continued)*

Application	Occupancy	X_{sm} (–)	$\dot{R}_{sm}$ (cig/$p_{sm}\cdot$h)	$\dot{V}_{vrp}/p$ (LPS/p)	$\dot{V}_{ets}/p$ Unadapted (LPS/p)	$\dot{V}_{ets}/p$ Adapted (LPS/p)	$\dot{V}_{tot}/p$ Unadapted (LPS/p)	$\dot{V}_{tot}/p$ Adapted (LPS/p)
Hospitality *(continued)*								
Bar, Cocktail Lounge (high)	100p/100 m²	0.5	2.0	5	28	20	32	24
Casino (low)	120p/100 m²	0.25	1.0	5	9	7	14	11
Casino (high)	120p/100 m²	0.5	2.0	5	28	20	32	24
Smoking Lounge								
(low)	70p/100 m²	1.0	3.0	3	33	26	36	29
(high)	70p/100 m²	1.0	6.0	3	66	51	69	55
Prison Cell	2 p	0.5	1.0	6	14	10	20	16

Table 6.1b. Summary of Some Examples and Additional Calculations on a Per Person Basis using the ETS Dilution Method with Specific Assumptions (SI Units for flow rates; rounded from I-P equivalents) *(continued)*

Application	Occupancy	X_{sm} (–)	$\dot{R}_{sm}$ (cig/p_{sm}·h)	$\dot{V}_{vrp}/p$ (LPS/p)	V_{ets}/p Unadapted (LPS/p)	Adapted (LPS/p)	V_{tot}/p Unadapted (LPS/p)	Adapted (LPS/p)
Hotel or Dorm								
Bedroom	2 p	0.5	0.6	7	8	6	16	13
Apartment or Condo Unit								
1 Bedroom	2 p	0.5	0.6	8	8	6	16	14
2 Bedrooms	3 p	0.67	0.6	8	9	6	17	14

7

SUMMARY

Environmental tobacco smoke, also known as secondhand smoke, is a controversial reality that we sometimes must address in our HVAC design work. Many decades of research and debate have occurred on this and closely related IAQ topics, and this long-awaited book will now, hopefully, to a large degree fulfill the need for tobacco smoke odor and irritation control information. More work remains to be done, but hopefully the technology and methods related to this topic are now sufficient for comfort design purposes.

As ETS is such a sensitive topic for some, including me, it may be a natural reaction to avoid related work. Through work by others, the standards of practice for ETS odor and irritation control are available, but some refinements will likely be made in the future. You should fully inform your clients as to the assumptions and limitations of designing for ETS, and make clear that *health issues cannot be addressed—only odor and irritation control can be attempted.*

Various methods for determining ETS-related ventilation air-flow rates exist, and several are presented in this book. One method presented in this book, developed for design purposes by others in the 1980s and 1990s using research results and firsthand experiences, seems to give reasonable rates that are purported by others to produce at least 80% acceptance of ETS odors and irritants, noting that the control of health risks is not its objective. The EDM and its factors, as with other HVAC design procedures, will likely continue to evolve. To help you understand and then use the EDM, many sample calculations are provided in this book, and Table 6.1 gives some results on a per person basis.

171

Ventilation should not be the only ETS engineering, architectural, or policy measure considered. Source control is by far the most effective way to reduce or eliminate an indoor contaminant; *a smoke-free building may be the best option in most situations.* Separation is then next, but the area that still has the secondhand smoke will need significant control measures. Providing properly designed smoking lounges is an example of separation that has enjoyed a high degree of success in some buildings. Local exhaust of the contaminant, and suitable makeup air, is an option for minimizing the mixing of a contaminant into the surrounding air, but the sources' locations typically must be fixed and thus is problematic with ETS. Dilution with ventilation air, and removal of "used" air, unfortunately does nothing to minimize the production of the contaminant, but it is often the most realistic approach for handling ETS's odors and irritants in many types of occupancies. Using a degree of displacement airflow, especially from floor to ceiling, can improve the contaminant removal effectiveness significantly. Air cleaning can be used, with care, to reduce the amount of outdoor air. This reduction typically saves energy and improves thermal comfort. Heat recovery, between the exhaust air and the outdoor air, can also significantly reduce energy use and/or be used to increase the ventilation rate. If smoking is to be allowed in a building or a particular space, it is likely that a combination of some or all of the preceding control methods will yield the optimal results.

As you prepare your designs, be sure to consider the seemingly secondary aspects as well, such as the ability of your systems to be adjusted or adapted to meet unexpected short-term conditions, or changed to meet long-term needs in the future. Also, making sure that your systems are relatively easy to operate and maintain will enhance, but not ensure, the likelihood that they will achieve their performance objectives over the long run. Legal issues and insurance coverage, for both the designers and owners, need full investigation with your advisers so that in the event of a design failure or claim, for example, responsibilities are clearly defined.

As with any area of practice, experience is invaluable. Be sure to observe how your and others' designs perform, and adjust your methods and choices so that future projects will be even more successful. Due to time and cost pressures, this feedback loop is often neglected, but if done consistently in your firm it should yield long-term benefits to you and your clients. Three known areas where enhancements are needed are: how to more accurately predict the ventilation effectivenesses; the expansion and refinement of data on percentage of smokers, smoking rates, and occupancy estimates for various ETS applications; and more specific dilution rates. When possible, be sure that you share your ETS design data and experiences widely so that others can benefit from your experiences as well—this sharing of knowledge is what makes us colleagues rather than competitors, and is expected in any profession.

Your suggestions for future editions of this book and related research and standards are encouraged. A form for submitting suggestions and comments appears at the end of this book.

NOMENCLATURE

A = area, floor, or duct cross-sectional (ft^2 or m^2)

ACH = air changes per hour (1/h)

$ADPI$ = air diffusion performance index

AHU = air-handling unit

C = concentration (e.g., ppm, ppb, lbm/ft^3, or kg/m^3)

ca = recirculated air

$\dot{D}_{cig}$ = design smoking density ($cig/h \cdot ft^2$ or $cig/h \cdot m^2$)

DI = dilution index

ea = exhaust air

E_{cr} = contaminant removal effectiveness (dimensionless)

E_f = filter efficiency (dimensionless)

II = irritation index

ka = makeup air

la = relief air

$\dot{m}$ = mass flow rate (e.g., lbm/min or kg/s)

ma	=	mixed air
MAU	=	makeup air unit
oa	=	outside or outdoor air
P	=	number of people (p)
P	=	pressure or static head (e.g., in.w.g., psi, or Pa)
pa	=	primary air
$\dot{Q}$	=	heat rate (Btu/h or W)
ra	=	return air
ρ	=	density (rho; lbm/ft^3 or kg/m^3)
$\dot{R}_{sn}$	=	smoking rate per smoker ($cig/p_{sm}{\cdot}h$)
RTU	=	rooftop unit
sa	=	supply air
t	=	time (e.g., s, min, or h)
T	=	temperature (°F or °C)
V	=	volume (ft^3 or m^3)
$\dot{V}$	=	volumetric flow rate, also sometimes Q (ft^3/min [CFM] or l/s [LPS])
V_{cig}	=	dilution volume of ventilation air (ft^3/cig or m^3/cig)
X_{oa}	=	outside air fraction (dimensionless)
X_{sm}	=	fraction of people who are smokers (dimensionless)

BIBLIOGRAPHY

ACGIH. *Industrial Ventilation: A Manual of Recommended Practice.* 24th ed. Committee on Industrial Ventilation. Cincinnati, OH: American Conference of Governmental Industrial Hygienist, 2001.

Alevantis, L., J. Wagner, W. Fisk, D. Sullivan, D. Faulkner, L. Gundel, J. Waldman, and P. Flessel. "Designing for Smoking Rooms," *ASHRAE Journal.* (July 2003): 26–31.

AMCA. *AMCA Fan Application Manual.* Arlington Heights, IL: Air Movement and Control Association, 1990.

ANSI/AIHA. *American National Standards for Laboratory Ventilation.* Z9.5-1992. Fairfax, VA: American Industrial Hygiene Association, 1992.

ANSI/ASHRAE. *ANSI/ASHRAE Standard 129-1997 (RA 02), Measuring Air-Change Effectiveness.* Atlanta, GA: ASHRAE, Inc., 1997.

———. *ANSI/ASHRAE Standard 62-1999, Ventilation for Acceptable Indoor Air Quality.* Atlanta, GA: ASHRAE, Inc., 1999.

———. *ANSI/ASHRAE Standard 62-2001, Ventilation for Acceptable Indoor Air Quality* and its various adopted addenda. Atlanta, GA: ASHRAE, Inc., 2001.

———. *ANSI/ASHRAE Standard 62.1-2004, Ventilation for Acceptable Indoor Air Quality* (and any adopted addenda). Atlanta, GA: ASHRAE, Inc., 2004a.

————. *ASHRAE Standard 62.2-2004, Ventilation and Acceptable Indoor Air Quality in Low-Rise Residential Buildings* (and any adopted addenda). Atlanta, GA: ASHRAE, Inc., 2004b.

————. *ANSI/ASHRAE Standard 55-2004, Thermal Environmental Conditions for Human Occupancy*. Atlanta, GA: ASHRAE, Inc., 2004c.

ASHRAE. *ASHRAE Handbook, HVAC Applications* vol. Atlanta, GA: ASHRAE, Inc., 1991.

————. *Design of Smoke Management Systems*. Atlanta, GA: ASHRAE, Inc., 1992.

————. *Air-Conditioning Systems Design Manual*. Atlanta, GA: ASHRAE, Inc., 1993.

————. *ASHRAE Handbook, HVAC Applications* vol. Atlanta, GA: ASHRAE, Inc., 1995.

————. *Psychrometrics: Theory and Practice*. Atlanta, GA: ASHRAE, Inc., 1996.

————. *ASHRAE Handbook, Fundamentals* vol. Atlanta, GA: ASHRAE, Inc., 1997.

————. *ASHRAE Handbook, HVAC Applications* vol. Atlanta, GA: ASHRAE, Inc., 1999.

————. *ASHRAE Handbook, Fundamentals* vol. Atlanta, GA: ASHRAE, Inc., 2001.

————. *Laboratory Design Guide*. Atlanta: ASHRAE, Inc., 2002.

————. *ASHRAE Handbook, HVAC Applications* vol. Atlanta, GA: ASHRAE, Inc., 2003a.

————. *IAQ Applications* vol. 4, no. 3 (Summer 2003), 2003b.

————. *ASHRAE Handbook, HVAC Systems and Equipment* vol. Atlanta, GA: ASHRAE, Inc., 2004.

Bauman, F. *Underfloor Air Distribution (UFAD) Design Guide*. Atlanta, GA: ASHRAE, Inc., 2003

Besant, R. W., and A. B. Johnson. "Reducing Energy Costs Using Run-Around Systems." *ASHRAE Journal* (February 1995): 41–46.

Bluyssen, P. M., and H. J. M. Cornelissen. "Addition of Sensory Pollution Loads—Simple or Not, That Is the Question." In *Proceedings of Healthy Buildings/IAQ '97* 2 (1999): 213–218.

Bohanon, H. R., Jr., and P. R. Nelson. "A Method for Measuring Air Cleaner Effectiveness." In *Procedings of the 8th International Conference on Indoor Air Quality and Climate* 2 (1999): 679–684, Edinburg, Scotland.

Bohanon, H. R., Jr., P. R. Nelson, and R. K. Wilson. "Design for Smoking Areas: Part 2—Applications." *ASHRAE Transactions* 104 (1998): 460–471.

Bohanon, H. R., Jr., J. J. Piade, M. Schorp, and Y. Saint-Jalm. "An International Survey of Indoor Air Quality, Ventilation, and Smoking Activity in Restaurants: A Pilot Study." *Journal of Exposure Analysis and Environmental Epidemiology* (2003): 378–392.

Brennan, P., P. A. Buffler, P. Reynolds, A. H. Wu, H. E. Wichmann, A. Agudo, G. Pershagen, K. H. Jöckel, S. Benhamou, R. S. Greenberg, F. Merletti, C. Winck, E. T. H. Fontham, M. Kreuzer, S. C. Darby, F. Forastiere, L. Simonato, and P. Boffetta. "Secondhand Smoke Exposure in Adulthood and Risk of Lung Cancer Among Never Smokers: A Pooled Analysis of Two Large Studies." *International Journal of Cancer* 109 (2004): 125–131.

Cain, W. S., B. P. Leaderer, R. Isseroff, L. G. Berglund, R. J. Huey, E. D. Lipsitt, and D. Perlman. "Ventilation Requirements in Buildings—I. Control of Occupancy Odor and Tobacco Smoke Odor." *Atmospheric Environment* 17 (1983): 1183–1197.

Cains, T., S. Cannata, R. Poulos, M. J. Ferson, and B. W. Stewart. "Designated 'No Smoking' Areas Provide from Partial to No Protection from Environmental Tobacco Smoke." *Tobacco Control* 13 (2004): 17–22.

CDC. "Strategies for Reducing Exposure to Environmental Tobacco Smoke, Increasing Tobacco-Use Cessation, and Reducing Initiation in Communities and Health-Care Systems." *Morbidity and Mortality Weekly Report* 49 (2000): 1–11.

―――. "Annual Smoking-Attributable Mortality, Years of Potential Life Lost, and Economic Costs—United States, 1995–1999." *Morbidity and Mortality Weekly Report* (2002): 300–303.

Chen, Q., and L. Glicksman. *System Performance Evaluation and Design Guidelines for Displacement Ventilation*. Atlanta, GA: ASHRAE, Inc., 2003.

DHHS. *Report on Carinogens*, 10th ed. Public Health Service, National Toxicology Program. Washington, DC: U.S. Department of Health and Human Services, 2002.

Dhital, P., R. W. Besant, and G. J Schoenau. "Integrating Run-Around Heat Exchanger Systems Into the Design of Large Office Buildings." *ASHRAE Transactions* 101 (1995): 979–991.

Doll, R., R. Peto, J. Boreham, and I. Sutherland. "Mortality in Relation to Smoking: 50 Years' Observations on Male British Doctors." *British Medical Journal* 328 (2004): 1519.

Dols, W. S., and G. N. Walton. *CONTAMW 2.0 User Manual*. NISTIR 6921. Gaithersburg, MD: National Institute of Standards and Technology, 2002.

El-Wakil, M. M. *Powerplant Technology*. New York: McGraw-Hill, 1984.

Enbom, S., I. Kumala, and A. Saamanen. "Reduction of Environmental Tobacco Smoke Dispersion in a Two-Compartment Restaurant." In *Proceedings of Healthy Buildings 2000* 2 (2000): 61–65.

Enstrom, J. E., and G. C. Kabat. "Environmental Tobacco Smoke and Tobacco Related Mortality in a Prospective Study of Californians, 1960–1998." *British Medical Journal* 326 (May 17, 2003).

EPA. *Respiratory Health Effects of Passive Smoking: Lung Cancer and Other Disorders*. Pub. No. EPA/600/6-90/006F. Washington, DC: U.S. Environmental Protection Agency, 1992.

Fanger, P. *Thermal Comfort*. New York: McGraw-Hill, 1972.

Gately, I. *Tobacco, the Story of How Tobacco Seduced the World*. New York: Grove Press, 2001.

Glantz, S. A., and S. Schick. Implications of ASHRAE's Guidance on Ventilation for Smoking Permitted Areas. *ASHRAE Journal* 46 (2004): 54–60.

Harriman, L., G. Brundrett, and R. Kittler. *Humidity Control Design Guide for Commercial and Institutional Buildings*. Atlanta, GA: ASHRAE, Inc., 2001.

Hyvarinen, M., K. Hagstrom, I. Gronvall, and P. Hynynen. "Reducing Bartenders' Exposure to ETS by Local Ventilation—Field Evaluation of the Solution." In *Proceedings of Indoor Air 2002* (2002).

IARC. *Tobacco Smoke and Involuntary Smoking*. Monograph series vol. 83. Lyons, France: Working Group on the Evaluation of Carcinogenic Risks to Humans, International Agency for Research on Cancer, World Health Organization, 2002.

Jadud, M. A., and B. A. Rock. "Tobacco Smoking Policy and Indoor Air Quality: A Case Study." In *Energy and Buildings* 20 (1993): 143–150.

Jenkins, R. A., D. Finn, B. A. Tomkins, and M. P. Maskarinec. "Environmental Tobacco Smoke in the Nonsmoking Section of a Restaurant: A Case Study." *Regulatory Toxicology and Pharmacology* 34 (2001): 213–220.

Kirkpatrick, A., and J. Elleson. *Cold Air Distribution System Design Guide*. Atlanta, GA: ASHRAE, Inc., 1996.

Klote, J., and J. Milke. *Principles of Smoke Management Systems*. Atlanta, GA: ASHRAE, Inc., 2002.

Knutson, G. W. "Ventilation for Odor Control," *ASHRAE Journal* (April 2003): 36–41.

Kuo, K. *Principles of Combustion*. New York: John Wiley & Sons, 1986.

Leaderer, B. P., and W. S. Cain. "Air Quality in Buildings During Smoking and Nonsmoking Occupancy." *ASHRAE Transactions* 89 (1983): 601–613.

———, W. S. Cain, R. Isseroff, and L. G. Berglund. "Ventilation Requirements in Buildings—II. Particulate Matter and Carbon Monoxide from Cigarette Smoking." *Atmospheric Environment* 18 (1984): 99–106.

———, and S. K. Hammond. "Evaluation of Vapor Phase Nicotine and Respirable Suspended Particle Mass as Markers for Environmental Tobacco Smoke." *Environmental Science and Technology* 24 (June 1991): 908–912.

Liu, R. T., R. R. Raber, and H. S. Yu. "Filter Selection on an Engineering Basis." *Heating, Piping, and Air Conditioning* (May 1991): 37–44.

McGinnis, J. M., and W. H. Foege. "Actual Causes of Death in the United States." *Journal of the American Medical Association* 270 (November 1993): 2207–2212.

Mckdad A., J. Marks, D. Stroup, and J. Gerberding. "Actual Causes of Death in the United States, 2000." *Journal of the American Medical Association* 291 (March 2004): 1238–45.

Morton, R. J. *Engineering Law, Design Liability, and Professional Ethics.* Belmont, CA: Professional Publications, Inc., 1983.

Muller, C., and W. England. "Achieving Your Indoor Air Quality Goals: Which Filtration System Works Best?" *ASHRAE Journal* 37 (February 1995): 24–32.

———, and R. Henriksson. "Proper Design and Use of Gas-Phase Air Filtration Systems for the Control of Environmental Tobacco Smoke." In *Proceedings of Healthy Buildings 2000* 2 (2000): 89–94.

Mumma, S. A., and K. M. Shank. "Achieving Dry Outside Air in an Energy Efficient Manner." *ASHRAE Transactions* 107 (2001): 553–561.

NAFA. *NAFA Guide to Air Filtration.* 2nd ed. Virginia Beach, VA: National Air Filtration Association, 1996.

Nazaroff, W. W., and B. C. Singer. "Inhalation of Hazardous Air Pollutants from Environmental Tobacco Smoke in U.S. Residences." In *Proceedings of Indoor Air 2002* (2002): 477–482.

Nelson, P. R., H. R. Bohanon, Jr., and J. C. Walker. "Design for Smoking Areas: Part 1—Fundamentals." *ASHRAE Transactions* 104 (1998): 448–459.

———, H. R. Bohanon, Jr., R. K. Wilson, F. W. Conrad, B. Mickens, W. D. Taylor, M. A. Huza, and G. H. Cox. "In Situ Measurement of Air Cleaner Ventilation Effectiveness." In *Procedings of the 8th International Conference on Indoor Air Quality and Climate* 4 (1999): 81–86.

NFPA. *NFPA Codes and Standards: NFPA 1 Fire Prevention Code; NFPA 13 Installation of Sprinkler Systems; NFPA 101 Life Safety Code.* Quincy, MA: National Fire Protection Association, n.d.

NIH. *Health Effects of Exposure to Environmental Tobacco Smoke.* Smoking and Tobacco Control Monograph no. 10, National Cancer

Institute, NIH Pub. No. 99-4645. Bethesda, MD: National Institutes of Health, 1999.

NSPE. *Code of Ethics for Engineers*. Alexandria, VA: National Society for Professional Engineers, 2003.

Pedersen, C. O., D. E. Fisher, J. D. Spitler, and R. J. Liesen. *Cooling and Heating Load Calculation Principles*. Atlanta, GA: ASHRAE, Inc., 1998.

Repace, J. L, W. R. Ott, and N. E. Klepis. *Indoor Air Pollution from Cigar Smoke*. Smoking and Tobacco Control Monograph no. 9, National Cancer Institute. Bethesda, MD: National Institutes of Health, 1998.

RJR. *Developing a Smoking Lounge*. Winston-Salem, NC: R. J. Reynolds Co., 1992.

Rock, B. A., and K. A. Moylan. *A Designer's Guide to Placement of Ventilation Air Intake Louvers*. (Final project report for ASHRAE Research Project [RP] 806.) Atlanta, GA: ASHRAE, Inc., 1998.

————, and D. Zhu. *Designer's Guide to Ceiling-Based Air Diffusion*. Atlanta, GA: ASHRAE, Inc., 2002.

SFPE. *The SFPE Handbook of Fire Protection Engineering*. 2nd ed. Quincy, MA: National Fire Protection Association, 1995.

SMACNA. *HVAC Systems—Duct Design*. 3rd ed. Chantilly, VA: Sheet Metal and Air Conditioning Contractors' National Association, 1990.

Steenland, K., K. Sieber, R. A. Etzel, T. Pechacek, and K. Maurer. "Exposure to Environmental Tobacco Smoke and Risk Factors for Heart Disease Among Never Smokers." *American Journal of Epideniology* 147 (1998): 932–939.

Straub, H. E., P. R. Nelson, and H. R. Toft. "Evaluation of Smoking Lounge Ventilation Designs." *ASHRAE Transactions* 99 (1993): 466–475.

Thayer, W. W. *Tobacco Smoke Dilution Recommendations for Comfortable Ventilation*. Paper 7092. Long Beach, CA: Douglas Aircraft Company, 1982.

U.S.S.G. *The Health Consequences of Smoking—A Report of the Surgeon General*. Washington, DC: CDCP Office on Smoking and Health, U.S. Department of Health and Human Services, 2004.

Vineis P., L. Airoldi, P. Veglia, L. Olgiati, R. Pastorelli, H. Autrup, A. Dunning, S. Garte, E. Gormally, P. Hainaut, C. Malaveille, G. Matullo, M. Peluso, K. Overvad, A. Tjonneland, F. Clavel-Chapelon, H. Boeing, V. Krogh, D. Palli, S. Panico, R. Tumino, B. Bueno-De-Mesquita, P. Feeters, G. Berglund, G. Hallmans, R. Saracci, and E. Riboli. "Environmental Tobacco Smoke and Risk of Respiratory Cancer and Chronic Obstructive Pulmonary Disease in Former Smokers and Never Smokers in the EIPC Prospective Study." *British Medical Journal* 330 (2005): 227–231.

Walker, J., P. R. Nelson, W. S. Cain, M. J. Utell, M. B. Joyce, W. T. Morgan, T. J. Steichen, W. S. Pritchard, and M. W. Stancill. "Perceptional and Psychophysiological Responses of Nonsmokers to a Range of Environmental Tobacco Smoke Concentrations." *Indoor Air* 8 (1997): 173–188.

Whincup, P. H., J. A. Gilg, J. R. Emberson, M. J. Jarvis, C. Feyerabend, A. Bryant, M. Walker, and D. G. Cook. "Passive Smoking and Risk of Coronary Heart Disease and Stroke." *British Medical Journal* (June 2004).

WHO. "Report on WHO Meeting, EURO Reports, and Studies 103." *Indoor Air Quality Research* (August 1986).

———. *The 2002 World Health Report*. Geneva, Switzerland: World Health Organization, 2002.

Whyte, W. *Cleanroom Design*. New York: John Wiley & Sons, 1999.

Wilson, D. J. "Flow Patterns Over Flat Roofed Buildings and Application to Exhaust Stack Design." *ASHRAE Transactions* 85 (1979): 284–295.

ABOUT THE AUTHOR

Brian A. Rock, Ph.D., P.E., Fellow ASHRAE, is an associate professor in the Civil, Environmental, and Architectural Engineering Department at the University of Kansas in Lawrence, Kansas. He received a B.S. in architectural engineering and a Bachelor of Environmental Design from the University of Kansas, his M.S. in engineering from the University of Texas at Austin, and a Ph.D. in engineering from the University of Colorado at Boulder. He is a registered professional engineer in Kansas.

Dr. Rock teaches HVAC system design to engineering students, and he performs research in many areas including indoor air quality, ventilation, and energy conservation. He previously did some writings on ventilation for ETS, and readily discusses the related engineering, health, and political issues with those from all points of view. Rock is a life-long non-smoker; he personally recommends not smoking.

INDEX

A

Acceptable IAQ, 24
Acoustics, 23–24
Activated carbon, 91
Active exposure, 14, 15
  body absorption, 15
  defined, 14
  See also Exposures
Adapted building users, 41
Additivity, 55–56
Adjustability, 98–100
  defined, 99
  fans, 99–100
  techniques, 99
  See also Design issues
Aerosols, 18
Air
  defined, 24
  exhaust (EA), 25, 28–29, 46,
    98, 133
  jets of, 34
  makeup (KA), 25, 133–34
  mixed (MA), 25, 75
  moist, 22

  outside (OA), 24
  primary (PA), 27
  recirculated (CA), 25, 64–65
  relief (LA), 25
  return (RA), 25, 33
  from smoking-allowed spaces,
    64
  supply (SA), 25, 83, 85, 98, 125
  transfer, 31, 63–64, 118–19,
    136–37
  ventilation, 31–32
  in "well-mixed" space, 34
Airborne pollutants, 17–20
  control, 19–20
  irritations, 18
  odors, 18
  particulates, 17–18
  RSPs, 18
  sensing, 19–20
  sinks, 19
  sources, 18–19
  VOCs, 17
Air change
  effectiveness, 37
  per hour (ACH), 158

Air cleaners, effectiveness, 124
Air cleaning, 38, 85–93
  defined, 69
  illustrated, 68
  references, 38
  removing particles, 86–91
  See also Contaminant control
*Air-Conditioning Systems Design
    Manual,* 6
Aircraft ventilation, 43, 49–54
  air bleed off, 49
  ceiling-to-floor air diffusion,
      49
  current recommendations, 54
  dilution index (DI), 50, 51
  floor-to-ceiling air diffusion, 49
  irritation index (II), 50
  for mixed seating, 53
  procedure, 50–52
  sample calculations, 52–54
  See also Ventilation air-flow
      rates
Air curtains, 75
Air diffusion performance index
    (ADPI), 38
Air distribution systems, 26–29
  defined, 26
  dual-duct, 26
  makeup air units (MAUs), 26
  packaged-terminal air
      conditioners (PTACs), 26
  packaged-terminal heat pumps
      (PTHPs), 26
  primary/secondary, 27–28
  readmission of exhaust air,
      28–29
  rooftop units (RTUs), 26
  single-duct, 26
  underfloor (UFAD), 5, 36
  unit ventilators, 26

Air exchange, 29–33
  defined, 29
  outside-air fraction, 32–33
  ventilation air, 31–32
Air-handling units (AHUs),
      24–25
  filter (F), 88, 89
  final filter (FF), 88
  heating capacity, 162
  illustrated, 25, 88
  local, 28
  outside air being admitted to,
      162–63
  prefilter (PF), 88
Air intake, placement, 13
Air lock, 74
Air outlets
  location, 84
  one-way, 85
Air pressure differentials, 30
Air retarders, 72
Air supply, in-floor, 5
Air-to-air heat recovery units, 32
Air transfer
  reducing, 96
  sliding/pocket doors and, 74
Air velocities, 96–98
  average, 98
  through open entryways,
      96–98
Air volume per cigarette, 56–59
  higher, 57
  lower, 57
  for mixed occupancies, 57–59
  ventilation air requirement, 58
Aldehydes, 11
Alkenes, 11
Americans with Disabilities Act
    (ADA), 113
Annoyance complaints, 77

ANSI/ASHRAE Standard 55,
    Thermal Environmental
    Conditions for Human
    Occupancy, 22–23
ANSI/ASHRAE Standard 55-2004,
    Thermal Environmental
    Conditions for Human
    Occupancy, 5
ANSI/ASHRAE Standard 62.1-2004,
    Ventilation for Acceptable
    Indoor Air Quality, 4–5
Apartments/condos, 157–61
    adjusted volume of dilution air
        per cigarette, 159
    example, 159–61
    high-rise, 157
    management control, 158
    per person rate, 167, 170
    restaurants/retail spaces, 158
    total ventilation air-flow rate, 160
    ventilation air-flow rate
        determination, 158
    See also Applications
Applications, 117–70
    apartments, condominiums,
        dormitories, 157–61
    bars, clubs, cocktail lounges,
        136–40
    casinos, 140–44
    hospitality, 132–44
    hotels/motels, 151–57
    offices, 117–32
    prisons, 149–51
    smoking lounges, 144–49
    support spaces, 161
Aromatic hydrocarbons, 11
Arrestance, 87
ASHRAE
    Air-Conditioning Systems
        Design Manual, 6

Cold Air Distribution System
    Design Guide, 5
Designer's Guide to Ceiling
    Based Air Diffusion, 5
design guide for underfloor air
    distribution, 5
duct design manuals, 5
Humidity Control Design Guide
    for Commercial and
    Institutional Buildings, 99
Principles of Smoke
    Management Systems, 6
Psychrometrics: Theory and
    Practice, 22
ASHRAE Handbook
    air cleaning, 38
    air intake and exhaust design,
        80
    defined, 4
    exhaust air reentry, 28
    fans chapter, 4
    fan sizing, 99
    flow rates, 48
    heat transfer, 109
    kitchen ventilation, 133
    local exhaust, 78
    psychrometrics, 22
    thermal comfort and acoustics,
        24
    thermal load calculations, 32
    ventilation and infiltration, 95
    ventilation rates in aircraft, 49,
        52, 53
ASHRAE Journal, 115
ASHRAE Load Calculation
    Manual, 6
ASHRAE Standard 62, 38–41
    development, 38–39
    IAQ procedure, 40
    parts, 39–40

ASHRAE Standard 62 *(cont'd.)*
  required ventilation air-flow
    rates, 39
  revisions, 39
  ventilation rate procedure, 40
Ashtrays, 12, 77
Audience, this book, 1
Automatic closers, 74, 96
Automatic controls, 101
Automatic sprinklers, 112

B

Backflow damper-fitted fans, 99
Barriers, 72
Bars, 136–40
  adjusted volume of dilution air
    per cigarette, 137
  contaminant removal
    effectiveness, 140
  examples, 137–40
  fixed seating, 136
  floor-to-ceiling displacement,
    136
  large, 136
  occupancy percentage, 137
  per person rate, 165–66, 168–69
  smoke-free, 136
  total occupancy, 138
  total ventilation air-flow rate,
    138, 140
  transfer air, 136–37
  unadapted occupants, 138
  See also Hospitality applications
Bioeffluents, 12
Break-rooms. *See* Smoking lounges

C

Carbon dioxide sensor, 102

Carpet, 112
Casinos, 140–44
  areas, 140
  base ventilation air-rate flow,
    142
  contaminant removal
    effectiveness, 143
  cutting-edge demand, 140
  example, 142–44
  free movement of occupants,
    141
  gaming areas, 141
  as multipurpose buildings, 140
  occupancy percentage, 142
  per person rate, 166, 169
  room air volume, 141
  slot machines, 141
  table games, 141
  ventilation rate procedure
    (VRP), 141
  See also Hospitality applications
Ceiling fans, 85
Challenge droplets, 88
Chewing tobacco, use of, 8
Cigarettes
  equivalent, 9
  size, weight, composition, 9
  tobacco, 8
  See also Smoke
Cigars
  smoke, 12
  tobacco, 8
Circulating fans, 85
Clubs. *See* Bars
Cocktail lounges. *See* Bars
Coils, 26
*Cold Air Distribution System
  Design Guide,* 5
Collectors, 90
Combustion, 10

Commissioning, 111
Commissioning agents, 32
Complaints, 129
Computational fluid dynamics
    (CFD) programs, 38
Concentrations, 19, 85
  contaminant, 20
  inlet, 46
  of materials, 45
  in perfectly mixed room, 46
  of RSPs, 48
  steady-state, 46–47, 123
  tracked, 45
Condominiums. *See* Apartments/
    condos
Conference rooms, 128–32
  HVAC system design for,
    128–29
  local exhaust, 130
  perfect mixing, 130
  per person rate, 165, 168
  table, 129
  total ventilation air-flow rate,
    131
  See also Offices
Connecting doors, 72–76
  air curtains, 75
  automatic closers, 74
  pocket, 74
  sliding, 74
  swinging and rotating, 73
  vestibule, 74
  See also Separation
Conservation equations, 44–49
  arrangement, 47
  concentration of exhaust air, 46
  conservation of species, 45
  mass flow rates, 44
  sample flow-rate calculation,
    47–49

steady-state room air
    concentration, 46
Constant air volume (CAV)
    systems, 32
Contaminant control
  air cleaning, 69, 85–93
  approaches illustration,
    68
  dilution, 68, 83–85
  local exhaust, 67, 77–82
  methods, 67–69
  separation, 67, 70–77
  source control, 67, 69–70
Contaminant removal
    effectiveness, 37, 56,
    122, 127
  bars, 140
  casinos, 143
  restaurants, 131
Contaminants
  concentrations, 20
  gas-phase, 91
  internal generation, 46
  outdoor air, 19
Continuous ceiling/floor plenums,
    31
Contractual release, 115
Controls, 101–3
  automatic, 101
  DCV, 101–3
  on/off, 101
Control volumes, conservation of
    mass, 44
Cooling
  energy consumption, 100
  evaporative, 33
  load calculations, 6
  process, 100
  See also Heating
Cost/benefit ratio (C/R), 114

Costs
  construction, 128
  HVAC system first, 23
  life-cycle (LCC), 114
  operating, 128

D

Decks, 26
Dedicated outdoor air system
      (DOAS), 28
Dehumidification process, 100
Demand controlled ventilation
      (DCV), 33, 101–3
  defined, 101
  exhaust systems, 103
  occupancy sensors, 101–2
  shut-down, 102
Depressurization, 93
  achieving, 94
  hotels, 154
  See also Pressurization
*Designer's Guide to Ceiling Based
      Air Diffusion,* 5
Design issues, 67–115
  adjustability, 98–100
  air cleaning, 85–93
  commissioning, 111–12
  contaminant control methods,
      67–69
  dilution, 83–85
  economic analyses, 114
  energy conservation,
      100–110
  engineering ethics, 114–15
  local exhaust, 77–82
  O&M, 111–14
  pressurization, 93–98
  separation, 70–77
  source control, 69–70

Design smoking density, 60
  hotels/motels, 156
  prisons, 150
  smoking lounges, 148
Differential pressure gauges, 95
Diffusers, 3
Dilution, 83–85
  adjusted volume per cigarette,
      122, 130, 137
  as contaminant control, 68
  displacement flow helps, 83–84
  illustrated, 68
  jets, 84–85
  in prisons, 149
  ratio, 29
  ventilation requirement, 123
  See also Contaminant control
Dilution index (DI), 50
Direct-contact heat exchangers,
      103–4
Direct exposure, 14
Displacement flow, 36
  ceiling-to-floor, 35
  defined, 35
  effectiveness and, 37
  floor-to-ceiling, 35, 36
  illustrated, 36
  use of, 172
Doors
  with automatic closures, 74, 96
  connecting, 72–76
  open, 31
  pocket, 74, 96
  sliding, 74, 96
  undercut, 31
  vestibule, 74
DOP Penetration Test, 88
Dormitories. *See* Apartments/
      condos
Dry scrubbing, 91

Dry-type extended surface filters, 88
Dual-duct systems, 26
Ducts, 31
   design manuals, 5
   exhaust, 81
   flexible, 37
   passages for, 76
   sizing, 100
Dust-holding capacity, 87

E

Economic analyses, 114
Electronic air cleaners (EACs), 88, 89–90
   collectors, 90
   defined, 89–90
   designs, 90
Electrostatic precipitators, 89
Energy conservation, 100–110
   controls, 101–3
   design, 100–101
   guidelines, 110
   heat recovery, 103–10
   measure (ECM), 64
   opportunity (ECO), 63
   See also Design issues
Energy recovery, 33
Engineering ethics, 114–15
Enhancements, areas needing, 172
Entrainment flow, 34
Entryways, 96
   air velocities, 96–98
   average air-flow velocity, 97
   depth, 96
   ETS-free side of, 97
   open, 31, 97
Environmental tobacco smoke.
   See ETS

*Environmental Tobacco Smoke
   Design Guide,* 49
Equivalent cigarettes, 9
Ethics, engineering, 114–15
ETS, 1
   absorption, 17
   acceptable levels, 10
   activity levels and, 22
   adsorption, 17
   airborne, 9
   air cleaner effectiveness, 124
   air treatment for, 38
   annoyance complaints, 77
   defined, 14
   design issues, 67–115
   in exhaust air, 28
   extra ventilation for, 56
   filter ratings, 87–88
   gas-phase contaminants, reducing, 91
   health consequences, 9–10
   odors/irritations, 10
   outdoor, 13
   plumes, 85
   presence of, 24
   scrubbing systems, 29
   spaces, identification, 23
   storage/removal processes, 16
   studying/designing for, 9
   thermal comfort and, 22–23
   ventilation rates, 41, 43–66
   zero concentration, 10
ETS areas
   air transfer/recirculation from, 85
   cleaned air and, 86
   cool operation, 101
   depressurizing, 93
   desired, list of, 115
   floor coverings, 112

ETS areas *(cont'd.)*
   paints, 112
   pressurization and, 93
   reuse, 114
   sealing, 100–101
   signage, 113
   ventilation air for, 85
   whole building, 96
   for working occupants, 101
ETS Dilution Method (EDM),
      54–61, 117
   additivity, 55–56
   adjustments, 66
   air volume per cigarette, 56–59
   base ventilation rate, 146
   defined, 54–55
   design smoking density, 60
   example, 61–65
   extra ventilation, 56
   intention, 55
   men versus women ratio and, 59
   results, 55
   smoking rates per smoker,
      59–60, 61
   solution, 61–65
   steady-state concentrations, 123
   See also Ventilation air-flow
      rates
ETS-free areas, 76
Evaporative cooling, 33
Exfiltration, 30
Exhaled smoke, 12
   composition, 16
   movement, 16
   See also ETS; smoke
Exhaust air (EA)
   balancing, 98
   carrying ETS, 28
   concentration, 46
   defined, 25

inlets/grills, 33
reentry, 28–29
restaurants, 133
See also Air
Exhaust fans, 79–80, 81
   air-flow rates, 79
   illustrated placement, 79
   smoking lounges, 145
   See also Fans
Exhaust grills, 145
Exhaust hoods, 78
Exhausts
   air-flow rate, 162
   DCV- controlled, 103
   ducts, 81
   hotel bathroom, 155
   industrial stacks design, 81
   intakes separation, 80–82
   placement, 80
   pressurization and, 94
   wall-mounted, 82
   wind direction and, 81
Exposures, 14, 19
   active, 14, 15
   passive, 14, 15
   predicting, 15

F

Fans
   coils, 28
   control, 99
   exhaust, 79–80, 81, 145
   reference, 4
   user control, 99
   variable capacity, 99
Feedback loop, 172
Filtered recirculated air, 64–65
   flow rate, 64
   no transfer air and, 65

Filter efficiency
  defined, 86
  determination, 65
Filters
  activated carbon, 91
  air, 86
  arrestance, 87
  cleanable, 86
  differential pressure sensors, 87
  dry-type extended surface, 88
  dust-holding capacity, 87
  ETS-removing, flow-rate, 92
  final, 89
  frames, 86
  furnace, 88
  gas-phase, 91
  HEPA, 89, 91
  high-efficiency, 89
  panel, 88, 89, 90
  particle, 88–91
  pleated and pocket, 88
  potassium permanganate, 91
  removable media, 88
  resistance to airflow, 87
  selecting, 92–93
  single, 88
  sizing, 92–93
  slots, 100
  super ultra low penetration air
      (SULPA), 89
  types of, 88–91
  ultra low penetration air
      (ULPA), 89
Filtration systems, 88
  panel filter, 88
  parallel fan-powered, 124
Final filters, 89
Fire egress, 96
Firsthand exposure, 14
Flexible ducts, 37

Floor coverings, 112
Fume scrubbers, 108
Furnace filters, 88
Future expansion, 100
Fuzzy wall coverings, 112

G

Gaming area, casinos, 141
Gases, 17
Gas-phase contaminants, 91
Gas-phase filters, 91

H

Health consequences, 9–10
  HVAC designers and, 10
  types of, 9
Heat exchangers (HXs), 26, 103
  air-to-air, 104
  counterflow, 103
  direct-contact, 103–4
  heat pipe, 108–10
  heat recovery, 109–10
  indirect, 104
  inspection provisions, 104
  parallel flow, 103
  plate-type, 104–5
  rotary, 105–7
  run-around, 107–8
  spray-type, 108, 109
Heating
  load calculations, 6
  process, 100
  sensible, capacity, 162, 163
  See also Cooling
Heating, ventilating, and air-
      conditioning. *See* HVAC
Heat pipe heat exchangers, 108–10
  defined, 108

Heat pipe heat exchangers *(cont'd.)*
  illustrated, 110
  optimal performance, 108–9
  See also Heat exchangers
Heat recovery, 103–10, 172
  effective, 162
  heat pipe heat exchangers,
      108–10
  plate-type heat exchangers,
      104–5
  rotary heat exchangers, 105–7
  run-around heat exchangers,
      107–8
  spray-type heat exchangers, 108
  use of, 161–63
  See also Energy conservation
Heat wheels. *See* Rotary heat
      exchangers
Heterocyclics, 11
High-efficiency particulate air
      (HEPA) filters, 89
  mounting, 91
  recommendation, 89
  See also Filters
Hospitality applications, 132–44
  bars, clubs, cocktail lounges,
      136–40
  casinos, 140–44
  designing for, 132
  occupancies, 132
  restaurants, 133–36
  thermal loads, 132
  See also Applications
Hotels/motels, 151–57
  base ventilation air-flow rate,
      156
  bathroom exhaust, 155
  computer rooms, 155
  corridors, 154
  depressurization, 154

design smoking density, 156
  example, 156–57
  floor gaps, sealing, 154
  hospitality classification,
      152–53
  occupancy percentage, 156
  per person rate, 167, 170
  pressurization, 154, 155
  room illustration, 153
  room recirculating fans, 153
  smoking/nonsmoking floors,
      switching, 153–54
  smoking/nonsmoking rooms,
      152
  smoking wing, 152
  spaces, 151
  stack effect, 152, 154
  total ventilation air-flow rate,
      157
  See also Applications
Humidification process, 100
*Humidity Control Design Guide
  for Commercial and
  Institutional Buildings,* 99
HVAC designers, 1
  air pressure differential creation,
      30
  as diverse group, 2–3
  guidance, 12
  successful, 2
HVAC systems
  acceptable IAQ, 24
  design, 2–3
  design introduction, 6
  engineering consultants, 4
  first cost, 23
  for large buildings, 3
  O&M manuals, 4
  operation and maintenance,
      3–4

primary/secondary, 27–28
for residential buildings, 3
system complexity, 3

I

IAQ procedure
  defined, 40
  requirements, 40
Inch-pound (I-P) system, 2
Indemnity clause, 115
Indirect heat exchangers, 104
Indoor air quality (IAQ), 20,
    24–41
  acceptable, 24
  air and airflows, 24–29
  air cleaning, 38
  air exchange, 29–33
  complaints, 129
  focus, 24
  Procedure of Standard 62, 92
  room air diffusion, 33–37
  ventilation, air change,
      contaminant removal
      effectiveness, 37–38
Indoor air temperature, 162
Indoor environmental quality,
    21–42
  adapted versus unadapted,
      41–42
  air quality, 24–41
  occupant perception, 24
  thermal comfort, 21–24
Industrial stacks, 81
Infiltration
  defined, 30
  reference, 95
Inhaled smoke, 12
  absorbed, 16
  filtered, 16

Insurance coverage, 172
Intakes
  exhausts, separation,
      80–82
  wall-mounted, 82
International System of Units (SI),
    2
Irritation index (II), 50
Irritations
  acceptable levels, 135
  acceptance, 171
  control, 171
  defined, 18
  reactions to, 22
  smoking lounges, 144
  tolerance, 42
  See also Odors

J

Jets, 84–85

K

Ketones, 11

L

Latent load, 162
Leadership in Energy and
    Environmental Design
    (LEED) program, 110
Legal issues, 172
Life-cycle cost (LCC), 114
Loads
  latent, 162
  peak, 162
  peak, steady-state analysis, 163
  steady-state analysis, 163
  thermal, 94, 132, 154

Load-savings calculation, 161–63
Local exhaust, 77–82
    air makeup, 77–82
    conference rooms, 130
    defined, 67
    illustrated, 68
    offices, 118–19
    via vent hoods, 78
    See also Contaminant control

M

Mainstream smoke, 12
    defined, 12
    illustrated, 13
    See also Smoke
Makeup air (KA) defined, 25
    restaurants, 133–34
    See also Air
Makeup air units (MAUs), 26
Mass flow rates, 45
Mechanical ventilation
    air-to-air heat recovery units, 32
    defined, 30
    pressurization and, 94
Method of test (MOT), 87
Minimum efficiency reporting value (MERV), 87, 88
Mixed air (MA)
    defined, 25
    large quantities, 75
    See also Air
Mixing
    boxes, 28
    ETS removal before, 128
    perfect, 34, 127
    reduction, 36
Moist air, 22

N

National Air Filtration Association (NAFA), 92
Natural ventilation, 29
Net occupiable floor area, 56
Neutral pressure level (NPL), 152
Nicotine, 7–8
    defined, 7
    observation, 7–8
Nitrogen compounds, 11
Noise, 23
Nonbuilding applications, 54
Nonsmokers
    odors/irritants tolerance, 42
    smoking lounges and, 147
Nonsmoking zone, 52

O

Occupancy
    dormitories, 158
    estimated maximum, 60
    fraction who are smokers, 59
    hospitality applications, 132
    mixed, 57–59
    percentage estimation, 130, 134–35, 156
    restaurants, 134
    sensors, 101–2
    smoker-to-total ratio, 58
    smoking lounges, 149
    total, 121, 138
Odors
    absorbing, 91
    acceptable levels, 135
    acceptance, 171
    adapted to, 41
    control, 171
    defined, 18

reactions to, 22
smoking lounges, 144
tolerance, 42
types of, 18
See also Irritations
Offices, 117–32
building users, 118
building variations, 117–18
conference rooms, 128–32
local exhaust, 118–19
open floor plans, 119–24
single-person, 124–28
speculative basis, 118
transfer air, 118–19
variable air volume systems,
118
VAV for, 118–19
See also Applications
Official interpretations, 5
Olfactory fatigue, 41
On/off controls, 101
Open doors/entryways, 31, 97
Open floor plans, 119–24
defined, 119
examples, 120–24
per person rate, 165, 168
See also Offices
Operation and maintenance
(O&M), 3–4, 111–14
architectural measures, 112
design for, 172
manuals, 4
practices information, 113
procedures standardization,
112–13
Organization, this book, 2
Outdoor air
acceptable, available, 31
contaminants, 19
percentage, 32

quality, 24
temperature/humidity levels, 30
upstream flows, 25
Outdoor ETS, 13
Outdoor smoking areas, 76–77
air intakes and, 76
covered, 76–77
designation, 76
Outside air (OA), 24
flow rates, 43
fraction, 32–33
See also Air
Oxidants, 11

P

Packaged-terminal air conditioners
(PTACs), 26
Packaged-terminal heat pumps
(PTHPs), 26
Paints, 112
Panel filters, 88, 89
in filtration system, 88
furnace, 88
thick, 90
See also Filters
Particles, 11
charges, 90
removing, 86–91
Particulates, 17–18, 19
settling, 19
size, 17–18
See also Airborne pollutants
Passageways, throat, 97
Passive exposure, 14, 15
Peak ventilation loads, 162
Percentage outside air, 32
Perfect mixing, 34
assumption, 46
conference rooms, 130

Perfect mixing *(cont'd.)*
  defined, 34
  performance, 127
  theoretical rate with, 43, 44–49
Per person rates
  apartments/condos, 167, 170
  bars/cocktail lounges, 165–66,
      168–69
  casinos, 166, 169
  conference room, 165, 168
  example summary, 163–64
  hospitality, 165–66, 168–69
  hotels/motels, 167, 170
  office, 165, 168
  open floor plan, 165, 168
  prisons, 166, 169
  restaurants, 165, 168
  single-person office, 165, 168
  smoking lounges, 166, 169
Physical separation, 71–72
Pipes, tobacco for, 8
Plate-type heat exchangers, 104–5
  defined, 104
  effectiveness, 104
  for ETS applications, 105
  illustrated, 105
  See also Heat exchangers
Pleated and pocket filters, 88
Pocket doors, 74
  air transfer and, 74
  with automatic closers, 96
  See also Doors
Potassium permanganate filters, 91
Prefilters, 88, 89
Pressurization, 93–98, 123
  achieving, 94
  air velocities through open
      doorways, 96–98
  defined, 93
  differential, 94, 95

differential gauges, 95
exhaust and, 94
hotels, 154, 155
maintaining, 95
mechanical ventilation and, 94
negative, 94, 95
positive, 94
VAV systems, 98
See also Design issues
Primary air (PA), 27
*Principles of Smoke Management
    Systems,* 6
Prisons, 149–51
  base ventilation air-flow rate, 150
  design smoking density, 150
  dilution ventilation, 149
  example, 149–51
  overcrowding, 149
  per person rate, 166, 169
  smoker percentage, 149, 150
  total ventilation air-flow rate,
      151
  See also Applications
Professional engineers (P.E.), 3
Psychrometrics, 22
*Psychrometrics: Theory and
    Practice,* 22
Public theaters, 132

R

Readings, 4–6
  must-have, 4–5
  other, 6
  related, 5–6
Recirculated air (CA)
  cleaned, use of, 64–65
  defined, 25
  upstream flows, 25
  See also Air

Recirculating systems, 85
Reducing agents, 11
Reentry
  defined, 28
  exhaust air, 28–29
Related readings, 4–6
Relief air (LA), 25
Removable media filters, 88
Resistance airflow, 87
Respirable suspended particles
    (RSPs)
  defined, 18
  low concentration, 48
  prediction, 48
Restaurants, 133–36
  areas, 133
  contaminant removal
    effectiveness, 135
  example, 134–36
  exhaust air requirements, 133
  kitchen, 133
  makeup air, 133–34
  occupancy, 134
  occupancy percentage,
    134–35
  per person rate, 165, 168
  physical floor-to-ceiling barrier,
    133
  seating area, 133
  smoking breaks, 133
  total ventilation air-flow rate,
    135
  transfer air velocities, 134
  unadapted occupants and, 136
  ventilation improvements, 134
  See also Hospitality applications
Return air (RA)
  defined, 25
  inlets/grills, 33
  See also Air

Rooftop units (RTUs)
  defined, 26
  placement, 82
  for venting ETS, 82
Room air diffusion, 33–37
  defined, 33
  displacement flow, 35
  entrainment flow, 34–35
  obstructions and, 37
  perfect mixing, 34
  supply air outlets and, 33
  UFAD, 36
Rotary heat exchangers, 105–7
  capacity, 105
  defined, 105
  heat wheel media, 106
  illustrated, 106
  vented gap, 106
  See also Heat exchangers
Run-around heat exchangers, 107–8
  advantages, 108
  defined, 107
  heat transfer fluid, 107
  illustrated, 107
  three-way valve, 108
  See also Heat exchangers

S

Scrubbers, 108
Scrubbing, 28–29
  dry, 91
  processes, 28–29
Secondhand exposure, 14
Sensible heating capacity, 162, 163
Sensors
  carbon dioxide, 102
  occupancy, 101–2
  selecting, 102
  VOC, 102

Separation, 70–77
  barriers, 72
  connecting doors, 72–76
  defined, 67
  degrees of, 70–77
  effectiveness, 71
  ETS-free spaces in different
      buildings, 71
  illustrated, 68
  intakes and exhausts, 80–82
  last degree of, 72
  outdoor smoking areas, 76–77
  physical, 71–72
  smoking receptacles, 77
  See also Contaminant control
Shut-down DCV system, 102
Sidestream smoke, 12
  characteristics, 14
  defined, 12
  illustrated, 13
  movement, 16
  temperatures, 14
  See also Smoke
Signage, 113
Simple payback (SPB), 114
Single-duct system, 26
Single-person offices, 124–28
  ceiling plan, 125
  dilution ventilation, 125–26
  examples, 126–28
  per person rate, 165, 168
  separate ventilation system,
      125
  supply air, 125
  transfer air, 125
  See also Offices
Sliding doors, 74
  air transfer and, 74
  with automatic closures, 96
SMACNA duct design manuals, 5

Smoke
  absorbing, 91
  combustion products, 11
  dilution, 17
  exhaled, 12, 16
  exposures, 14
  flows, 12–14
  inhaled, 12, 16
  mainstream, 12, 13
  movement, 15–17
  production, 10–17
  residuals, 12
  sidestream, 12, 13
  traveling distance, 14
  See also ETS
Smoke-free bars, 136
Smoke-free buildings, 172
Smokers
  chain, 48
  insurance premiums, 9–10
  occupants fraction, 59
  odors/irritants tolerance, 42
  ratio determination, 57–58
  smoking rates, 59–60
Smoker-to-total occupancy ratio,
    58
Smoking lounges, 144–49
  base ventilation rate, 146
  break-rooms, 144
  conceptual design, 146
  design smoking density, 148
  displacement ventilation, 149
  exhaust fan, 145
  existing rooms, 145
  irritants control, 144
  nonsmoker assumption, 147
  occupancy, 149
  odor control, 144
  per person rate, 166, 169
  public interest, 144

raised floor, 145
retrofit, 145
success, 172
total ventilation air-flow rate,
    145, 148
usage rates, 145
ventilation air-flow rate per
    person, 148
See also Applications
Smoking rates, 122, 126, 131
Smoking rates per smoker, 59–60
defined, 59
table, 61
Smoking receptacles, 77
Smoldering, 10
Sound, 23
Source control, 69–70
defined, 67
design needs, 69–70
example, 69
illustrated, 68
See also Contaminant control
Spray-type heat exchangers
defined, 108
in fume scrubbers, 108
illustrated, 109
See also Heat exchangers
Stack effect
defined, 152
illustrated, 154
Subdivide open space, 70
Super ultra low penetration air
    (SULPA) filters, 89
Supply air (SA)
balancing, 98
defined, 25
flow rate variation, 98
grill, 85
injecting, 83
outlets, 33

single-person offices, 125
total flow rate, 32
See also Air
Support spaces, 161

T

Terminal units, 28
Terpenes, 11
Test, adjust, and balance (TAB)
    technicians, 32, 111
Theoretical rate with perfect
    mixing, 43, 44–49
conservation equations, 44–49
defined, 44
See also Perfect mixing;
        Ventilation air-flow rates
Thermal comfort, 21–24
acoustics, 23–24
defined, 21
ETS and, 22–23
maintaining, 36
thermal zoning and, 23
See also Indoor environmental
        quality
Thermal complaints, 129
Thermal loads
calculation, 32, 94
on hallways, 154
hospitality applications, 132
Thermal Comfort, 21
Thermal zones, 23, 118
Tobacco
chewing, 8
in cigarettes, 8
in cigars, 8
growth, 8
nicotine, 7–8
origin, 7
oxidation rate, 10

Tobacco *(cont'd.)*
  for pipes, 8
  plants, 7
  soil depletion characteristics, 8
  use, 8, 9–10
Total occupancy, 121, 138
Total ventilation air-flow rate, 63
  apartments/condos, 160
  bars, 138, 140
  conference rooms, 131
  hotels/motels, 157
  open floor plans, 121, 122
  prisons, 151
  restaurants, 135
  single-person offices, 127
  smoking lounges, 145, 148
  See also Ventilation air-flow
    rates
Transfer air
  bars, 136–37
  defined, 31
  injecting, 83
  offices, 118–19
  single-person offices, 125
  through doorways, 64
  use of, 63–64
  See also Air
Transfer grills, 31
Transoms, 31

U

Ultra low penetration air (ULPA)
    filters, 89
Unadapted users
  in bar calculation, 138
  defined, 41
  in restaurant calculation, 136
Undercut doors, 31

Underfloor air distribution
    (UFAD), 5, 36
Uniform Mechanical Code (UMC),
    28
Units, used in this book, 2
Unit ventilators, 26
Urns, 12
U.S. Green Buildings Council, 110

V

Vapor barriers, 72
Variable air volume (VAV), 33
  flow rate, 33
  office use, 118–19
  single-duct systems, 119
  systems, 98, 118
Ventilation
  air, 31–32
  in aircraft, 43
  air-flow rates, 39, 40
  common goal, 30–31
  defined, 29
  demand controlled (DCV), 33,
    101–3
  effectiveness, 37
  mechanical, 30, 32
  natural, 29
Ventilation air-flow rates, 43–66
  aircraft ETS, 49–54
  determination methods, 171
  ETS Dilution Method (EDM),
    54–65
  per person, 131
  studies, 43
  theoretical rate with perfect
    mixing, 43, 44–49
  very high, 47
  See also Total ventilation air-
    flow rate

Ventilation rate procedure
  casinos, 141
  defined, 40
Vestibule, 74
Vibrations, 23
Visitors, 41
Volatile organic compounds
        (VOCs)
  defined, 17
  sensors, 102

W

Wall-mounted intakes/exhausts, 82
Wind direction, 81
World Health Organization
        (WHO), 9, 48

COMMENT SHEET

VENTILATION FOR ENVIRONMENTAL TOBACCO SMOKE
FIRST EDITION
ELSEVIER INC. 2006

Readers are encouraged to submit their comments, corrections, and suggestions for future versions of this book. Please make and then fill-out a photocopy of this page for each comment you have. Mail or fax the completed page(s) and legal copies of any supporting documents to the author whose contact information can be found at http://engr.ku.edu/~brock/rock.htm.

Name: _____

Date: _____

Mailing Address: _____

_____

Telephone: _____

E-mail: _____

This comment is in regards to page _____, paragraph _____:

_____

_____

_____

_____

_____

_____

_____

_____

_____

_____

Thank you for your comments, corrections, or suggestions.